Science and Fiction

T0059869

Science and Fiction - A Springer Series

This collection of entertaining and thought-provoking books will appeal equally to science buffs, scientists and science-fiction fans. It was born out of the recognition that scientific discovery and the creation of plausible fictional scenarios are often two sides of the same coin. Each relies on an understanding of the way the world works, coupled with the imaginative ability to invent new or alternative explanations - and even other worlds. Authored by practicing scientists as well as writers of hard science fiction, these books explore and exploit the borderlands between accepted science and its fictional counterpart. Uncovering mutual influences, promoting fruitful interaction, narrating and analyzing fictional scenarios, together they serve as a reaction vessel for inspired new ideas in science, technology, and beyond.

Whether fiction, fact, or forever undecidable: the Springer Series "Science and Fiction" intends to go where no one has gone before!

Its largely non-technical books take several different approaches. Journey with their authors as they

- Indulge in science speculation - describing intriguing, plausible yet unproven ideas;
- Exploit science fiction for educational purposes and as a means of promoting critical thinking;
- Explore the interplay of science and science fiction - throughout the history of the genre and looking ahead;
- Delve into related topics including, but not limited to: science as a creative process, the limits of science, interplay of literature and knowledge;
- Tell fictional short stories built around well-defined scientific ideas, with a supplement summarizing the science underlying the plot.

Readers can look forward to a broad range of topics, as intriguing as they are important. Here just a few by way of illustration:

- Time travel, superluminal travel, wormholes, teleportation
- Extraterrestrial intelligence and alien civilizations
- Artificial intelligence, planetary brains, the universe as a computer, simulated worlds
- Non-anthropocentric viewpoints
- Synthetic biology, genetic engineering, developing nanotechnologies
- Eco/infrastructure/meteorite-impact disaster scenarios
- Future scenarios, transhumanism, posthumanism, intelligence explosion
- Virtual worlds, cyberspace dramas
- Consciousness and mind manipulation

More information about this series at http://www.springer.com/series/11657

Harun Šiljak

Murder on the Einstein Express and Other Stories

 Springer

Harun Šiljak
International Burch University Sarajevo
EEE Department
Sarajevo, Bosnia and Herzegovina

ISSN 2197-1188   ISSN 2197-1196   (electronic)
Science and Fiction
ISBN 978-3-319-29065-2   ISBN 978-3-319-29066-9   (eBook)
DOI 10.1007/978-3-319-29066-9

Library of Congress Control Number: 2016942550

Cover illustration: A train by night slow exposure. From CanStockPhoto.

Printed on acid-free paper

This Springer imprint is published by Springer Nature
The registered company is Springer International Publishing AG Switzerland

Preface

Murder on the Einstein Express is my first collection of stories, and it contains almost everything I wrote in the science fiction genre over the past 7 years. The four stories in this volume range from Alice in Wonderland in the real analysis setting (Normed Trek) over a computer-based mathematical proofs dystopia (The Cantor Trilogy) and Arabian nights from the future (In Search of Future Time) to a story about a hideous crime in an imaginary train, told in a Russian classroom (Murder on the Einstein Express).

Stories in this collection started coming to life in 2008 with Normed Trek, which was written right after my Calculus II final exam. Two years later, Murder on the Einstein Express was written: first as a flash fiction crime story and then as the nested narrative which I see as the backbone of this collection. Cantor Trilogy was written in 2014, and In Search of Future Time came a year later, first as a sequence of separate things, then as a parallel storyline in Murder on the Einstein Express, and finally as a separate piece.

It is hard to put a genre tag on the stories collected here, especially for Murder on the Einstein Express and Normed Trek. Both of them were primarily written without plans for publishing, simply to move them from the world of neurons to the world of ink, paper, and/or computer memory. As a result, I had a fictional story about science, and not science fiction: such story would hardly find its home outside of my drawer or PC or way to the readers without a book series like Springer's Science and Fiction.

This story collection is a mixture of science and fiction: most of the time the reader is able to separate one from the other, but sometimes the question "fact or fiction" remains unanswered. If there is a central idea in the book, it is David Hilbert's famous "Wir mussen wissen, Wir werden wissen" (We must know, we will know): the characters, real and imaginary, want knowledge, unbounded and complete. Every story hides a homage to real, wise, inspiring people the author admires and loves.

Most of the science fiction in the stories is focused around artificial intelligence, as the author's research areas overlap with the work of the scientific community in artificial intelligence. There is no clearly visible threat to humans in these stories – there is just concern for AI psyche. Caveat: invisible threats ahead.

I would like to thank a number of people here: those who inspired parts or whole stories, those who read it and gave comments and their support, and of course the team at Springer.

Sarajevo, Bosnia and Herzegovina
September 2015   Harun Šiljak

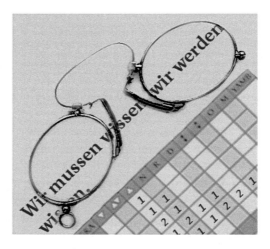

Hilbert's sentence, "Wir mussen wissen, wir werden wissen" (We must know, we will know), is in certain way the foundation of this book in which the pursuit of knowledge and question of our knowledge limitations are repeated in every story. Pince-nez glasses are David Hilbert's trademark. The form seen in the cover is used in a dice game popular in former Yugoslavia (and probably elsewhere), called Yamb (God doesn't throw dice, A. said). The numbers in the form are just one more puzzle for you: what is the next number in the sequence 1, 11, 21, 1211, 111221? (Image by the author)

Contents

Part I

First Story

Normed Trek[1]

Limit, the final frontier. These are the voyages of the function e^x. Its continuous mission: to explore new normed spaces… to find new norms and metrics… to boldly converge where no function converged before.

$t = 0$

At the moment of $t = 0$, we arrived in the space B. Our group (denoted G) was small, but commutative with respect to addition, composed of different non-negative elements. The trajectory that lead us to this space was class C^∞, smooth and continuous. The mission for the group remained unknown, although given implicitly.

It is the time to introduce myself. I am the supremum of the set of our group members, its maximum, actually… since the set is closed. My name in the epsilon-delta language was *The one equal to his derivative*. You may call me e^x. Strictly increasing, monotone, exponential. According to the D'Alembert's test, I was the best choice for this mission, the only one to fulfill the Lipschitz condition on the hyperrectangle given in the well-protected basis of vector space V. I had the right to select a countable number of elements in my set, which I did, via a bijection with the set of integers to the group. I chose the members wisely, by the Cauchy's test – its strongest form.

While the days uniformly converged to the day of our departure, periodic function of my heart started shrinking its period, excitement grew factorial-wise. But the mission still remained a secret, multiplied with Heaviside function, available to us only at $t = 0 + \varepsilon$ for an arbitrary small ε, of course.

[1] Originally published as Šiljak, H. "Normed Trek", The Mathematical Intelligencer, Volume 36 Issue 3 (September 2014), pages 53–55. DOI: 10.1007/s00283-014-9465-1. © 2014 Springer Science and Business Media New York.

I had to split with my family, the $e^x + C$. They kept our famous dignity while saying farewell. "Integrable (Riemann sense) even when it's hardest", one could say.

While $t \to 0_-$, the Dirichlet kernel started its action on me. I didn't know what transform is it going to perform, I just knew I might easily be not equivalent to what I once was. Shall I ever return to our Euclidean Space? Shall an inverse transform ever make me the same I am today? What awaits in space B?

$t = \frac{\pi}{2}$

I moved just an epsilon from the Origin, making insecure steps through the Space B. I was still the same exponential function, the transform did nothing to me. Thank Supremum for that!

I was afraid this was one of those spaces based on a set with cardinality $\aleph_0$. At one point in those spaces you're safe, but the next step can lead you to disappearance... that is why I preferred the spaces with the power of continuum: better uncountable than incomplete, as my father would say. Luckily, this space, as wild as it looked like, didn't cause problems for my movement: I felt like moving in the $[0, 1]$ segment, and not in a strange, distant space. I examined the sinusoids of mountains, diverging in the infinity and thinking about the $\mathbb{R}^2$ plane I grew up in. Alas, there wasn't time to think. Our mission has been officially stated: we are looking for an inverse Melin Transform of one of the functions from our Euclidean Space, which disappeared recently. Periodic functions have that habit – disappearing and coming back. This function, as the members of her class informed us, isn't periodic at all.

It hasn't been clear how are we going to do our task, especially knowing that there are a lot of members of this B-Space that don't like us being here. Ever since $t = 0$, I noticed that in an arbitrary small neighborhood of every one of us there had been countably infinite number of asymptotic followers: simply put, spies. Who sent them?

I knew that the ruler of this world is the infamous Dirichlet Function $\chi(x)$ (usually called just Function). Discontinuous at every point, graphically inexpressible, periodic without a period... Fascinating and frightening properties. Not many functions in our Euclidean Space lived their lives without knowing of this Function's existence (and uniqueness), and her everyday wrongdoings in the B-Space. Still, there was nothing we could do, but hope that the vector of life won't bring us to her domain.

As they say, life is a convolution of luck and its inverse – and I got majorised by the inverse. Obviously.

In one of the epsilon neighborhoods of the Origin we met a series with a curious general term $\sin nx/(n \ln n)$. He introduced himself as a Fourier series of one function not known well in our Space... and he offered services as a guide through trajectories of B-Space. Other members of the group didn't find this series strange, but I definitely didn't like his story. True, it was a trigonometric series, even uniformly convergent (I'm sure younger female members of my group liked that a lot, together with that modern stylish graph he had). Nevertheless, using the Parseval theorem (in V space we were reminded about it just before leaving), I have easily proven what I conjectured.

He is *not* a Fourier series.

Ignoring that for a moment, I let him guide us, with sole desire to see what is the Function up to. She obviously knew we are here.

$t = \pi$

The way which the alleged Fourier series chose looked more and more like a logarithmic spiral. We were moving, but I didn't see us approaching our goal, the Accumulation Point of B-Space (to be more precise, my goal was the Kernel of B-Space). Flocks of sinusoids flew over our level surface, disturbing the flux of my thoughts. I was thinking about danger that might be ahead in the trajectory given. My knowledge of weapons used here was fairly limited, because there weren't many survivors after their use. Weapons I was warned about were the Differentiator and Riemann Integrator. Principle of their action was pretty simple: making derivatives and definite integrals out of the functions aimed at, respectively. Integrator sounded deadly... and it was, because turning to a constant means death for functions from most spaces (luckily, there was only one Riemann Integrator in whole B-Space, held deep in its kernel, where only the Function and the most loyal Metric Guards have access to). There were stories of mass integrations and differentiations in integration and differentiation camps, respectively. Although this action has been forbidden by the Abel-Laplace theorems, there is already evidence of such procedures being conducted.

Beside these dangerous operators, functions in this space use multiplicators and additors in everyday life: operators for multiplication and addition with a non-zero real constant, respectively. Such actions are considered harmless. Such behavior was even encouraged by the Function herself!

There were also stories of terrible guards of the camps: fast exponentials and factorials you can't run away from. We could just hope we'll never get there.

When I informed the rest of the team of the fake Fourier series taking us Supremum-knows-where, they didn't believe at first. "By Supremum of All Sets", I swore, "if Parseval is lying, then so am I. But beware of this series. I have to stay behind – this essential left discontinuity we are approaching is something I have to investigate on my own."

The rest of the team continued the journey, keeping an eye on their guide, while I stayed in the essential discontinuity, looking for a clue to take me to the missing function.

$$t = \frac{3\pi}{2}$$

Function we were looking for was from an old, noble family of Lebesgue integrals. She was still an infinitely small function when she left the Space using the Inverse Melin Transform. It is still unknown who made it possible for her to conduct this dangerous transform, but there was a feeling that the infamous Function had its asymptotic fingers in this. Young and inexperienced function from the powerful house of Lebesgue Function most probably planned to use for blackmail of our Space's leaders.

This information was all I had at the beginning of my quest. By pure chance, just around this discontinuity where the night half-period of the eternal time sinusoid found me, I met a group of power functions that looked slightly odd. A casual conversation revealed that they are here on a special mission: conducting Direct Melin Transform of functions coming here from my own home Space! Then, as they say, these functions are sent to the kernel. What happens there, they couldn't say.

So, the function I'm after is in B-Space's kernel, and it's in the form I can recognise it in: $(x + \pi)^{-1}$. Moving for an epsilon from the discontinuity rising above the hyperplane we were in, I found a sequence that was converging rapidly towards the kernel. I joined them, asked for news: and the news were bad.

An agent of Metric Guards has lured a group of B-Space and Metric's hated enemies into a trap (obviously, it was my group). They were left in the integration camp "Jordan Measure Zero". If and when the integration happens, Function herself will do it.

Obviously, they weren't careful enough. I shouldn't have left them alone. But what good would it do if the Guards had caught me too?

Voice of a gamma function at the kernel entrance made a jump discontinuity in my thought process at that moment: "Welcome to the heart of B-Space. What brings you here?"

"Job advertisement. Guard job in "Jordan Measure Zero"… I replied.

I expected this question, of course, so while staying around that essential discontinuity, I read the job advertisements in yesterday's "Mobius Band" (*B*-Space daily newspaper), where I discovered that the Metric Guards look for exponential functions to work as guards in "Facility of function testing – Jordan Measure Zero" (what an euphemism for that horrible camp).

"Go to the Hamel basis of *B*-Space. Second derivative from the left", said the guard in an official tone and let me in the kernel.

$t = 2\pi$

Obviously, Guards needed new guards. They didn't interrogate me much, satisfied with the fact that an exponential function is willing to spend the rest of its days protecting *B*-Space from the enemies of the Metric. My job, as they said, starts in $\pi/2$ from now, when Function is going to the "Jordan Measure Zero" to attend the foreign spies' execution. When I asked why is this execution so important, they gave me a surprising answer: it will be the premiere of the new technological miracle of the Metric Guards, RAI. RAI stands for, as they explained, Riemann Absolute Integrator, an operator capable of turning even nonintegrable functions into a mere real constant, only if they are absolutely integrable.

(Luckily, it seems that taking functions not even absolutely integrable from the Weierstrass Squadron has been a good idea. The nondifferentiable, non-integrable rogues will be hard to neutralise… until they invent a multiplicator multiplying with zero).

Beside the Function, a function joining the kernel only recently is going on this trip as well, they said. Unknown to functions from this space, one with an elegant and seductive graph.

"By the Supremum, I've never seen a hyperbola that pretty", a polynomial screamed, in religious passion. "Vertical asymptote around the holy number, in the negative… Marvelous", he went on mesmerised.

Hmmm. Holy number… in the negative. $-\pi$. That has to be her, the function I'm looking for. So, I will see her in $\pi/2$. Until then, I have to come up with a rescue plan.

The idea was emerging. Integrating absolute values, functions in the camp bounded to a segment. Maybe, just maybe… the untouchable Function can be tamed.

$$t = \frac{5\pi}{2}$$

Hodograph toward the camp was ahead, as the time sinusoid passed its maximum for the second time since I entered the B-Space. Two Metric Guards, the Function, $(x + \pi)^{-1}$ and me were in the group moving down the hodograph. I can't describe the function, it can't be done graphically. I can just describe the feeling she caused for all of us: fear and shivers. There were no bodyguards present around the Function. No weapon, as the wise Metric Guards ensured her, could do any harm to such a Function.

Guards in the group were there because of $(x + \pi)^{-1}$. They probably didn't have enough trust in me yet, so they took care of young beauty's secure transport themselves. At this point, I could see for myself that the description of the beauty given to me in the kernel was correct. I knew a few of her relatives, members of family $(x + C)^{-1}$ for a real C, but none had such grace, such elegance in the graph. The very fact that $C = \pi$ for her (I'm not the only person seeing π as the most beautiful constant of the real axis) was enough to wake up some feelings inside me... those feelings I had no proof of existence (or uniqueness) before. Now, I had a constructive proof.

Is it possible that a function I'm seeing for the first time in my life has already solved the differential equation of my heart? Love is a function of infinitely many variables, but she did find my particular solution, solved my Cauchy's problem. Still, I didn't reveal my identity.

Pretending to be absentmindedly playing with the additor, I tested its action on fellow Guards. I noticed the functions don't really notice addition of a constant to them. Another puzzle block in my plan fits into its position. I pointed the additor at the Function and set the value up to $-1/2$. She didn't react, just like no one else did.

The general opinion of the public in the B-Space is that constants don't play a role in the everyday existence. Well, we'll see about that.

Sinusoid of time was approaching the abscissa as we entered the camp area.

$$t = 3\pi$$

At the entrance, they limited us all to the bounded segment $\langle a, b \rangle$. Even the Function had to be limited, like the prisoners, since those were the precaution measures made by Metric Guards. Two Guards who escorted us to the camp went to gather the prisoners, while Function, $(\pi + x)^{-1}$ and me were left alone. RAI was there, not protected in any way: clearly, the Function didn't suspect the danger that threatened. I grabbed it and pointed at her.

After a surprised and a furious look she didn't say anything, but I did. "I trade your life for the life of prisoners. If I start RAI, it will make a pathetic constant out of you."

"Are you stupid enough not to know that the absolute value of Dirichlet's function is equal to itself? You fool."

"But you're not $\chi(x)$ anymore, but $\chi(x) - 1/2$. And the absolute value of that function is $1/2$. You're smart enough to know that a constant is integrable on a bounded segment (we're all currently in one). Clear?"

"Clear", she admitted, "but what do you want for my life? What good for you does my death bring you?"

"No good for me, but I'm sure your servants would enjoy it… just as much the Guards would suffer. Although I would enjoy sending you in the Empty Set, I will not do it if you return all your prisoners, $(x + \pi)^{-1}$ and me in our Space at once. No tricks!"

It's always big, the fear of rulers. Prisoners were gathered at the initial point of a trajectory leading to the Origin. They waited for $(x + \pi)^{-1}$ and me. We went there, but I took the Function with us as well. She will be set free in the Origin, but until then she is our insurance.

Luckily, no problems were encountered during our trip and we soon found ourselves an arbitrarily small neighborhood of the Origin.

"I am sorry for not setting your people free instead of setting you free, but I'm sure that now, once you realised how vulnerable you are – you won't be that cruel to them", I said, observing the Function as I tossed the RAI in an essential discontinuity nearby. That discontinuity reminded me of my childhood and the great monument dedicated to division by zero in my hometown.

"Thank you for your chivalry and holding to your word… but don't expect the same noble deeds from me if we meet again", the Function hissed, running away.

I stood there, deep in my thoughts, looking (hopefully, for the last time) flocks of sinusoids flying above the B-Space's scalar fields. Two branches of a hyperbola hugged me tight: $(x + \pi)^{-1}$. She whispered: "I owe you my life. Can I share it with you?"

At $t = \frac{9\pi}{2}$, we arrived in the Space.

$t = 4\pi$

$\pi/2$ had passed after our return to the Space, but I had to see $(x + \pi)^{-1}$ again. I slowly walked through the direction field in front of her family's house (in the Space, we use the term stationary point, though) thinking what to say. I was carrying a bouquet of isoclines plucked in that very field. She approached me.

"We are too different. You're a rational function, and I'm an exponential. We are not meant to be together, our convolution would never be something our families would accept", I said, trying to bound the sinusoid of my excited voice.

"I know, that is why I have decided to take the inverse Laplace transform and become an exponential. I guess your family will accept that…" she whispered.

"But what about *your* family?", I asked, surprised.

"My mother was an exponential before marriage and took the Laplace transform to become a rational. She knows very well what a sacrifice for love is."

I didn't have anything to say after these wonderful words. I stared and stared at the charming foci of her hyperbola, trying to find out finally does that function, function that has become the only solution of my life's system of equations, wish to become asymptote of my heart:

"Do you want to convolve with me, $(x + \pi)^{-1}$?"

"Yes, e^x. Until infinity do us part."

Silence.

"$x^2 + 2(y - \frac{3}{4}\sqrt{|x|})^2 = 1$", I whispered.

Part II

Second Story

In loving memory of Abdus Salam Cantor, the first ⬛⬛⬛⬛⬛ Nobel laureate for its work in physics.

Cantor Trilogy[1]

Waiting for Pseudoscience

Previously unpublished letter found in Hastings Institute Museum inside one of the books owned by J. L. Hastings, signed by certain György Molnar. Footnotes and comments by Jennifer Misley, Hastings Curator.

Dear Professor Hastings,[2]

I am one of the undergraduate students in your Cantor Architecture[3] courses, and as most of my colleagues, I'm impressed with it. The whole concept of Cantor and the cantor networks is overwhelmingly impressive and surely unimaginable only a few decades ago.

As a math major, I have started reading the masters, as you always advise us to (seems like every course this semester can be learned from easily readable masters). It lead to an interesting thought experiment I would like to share with you, since it is directly related to the concept of Cantor and artificial intelligence in general, at least the intelligence aiming at helping us in scientific work and not the artificial intelligence science fiction writers are rooting for in their writings. I hope you will find it interesting, or – even better – prove me wrong and bring me faith in the future of cantors.

For the Algorithms course in Prof. Starr's class we read Grey's treatise on Turing machines and there is a wonderful chapter in it, dedicated to explaining

---

[1] Originally published as Šiljak, H. "The Cantor Trilogy," Journal of Humanistic Mathematics, Volume 5 Issue 1 (January 2015), pages 299–310. DOI: 10.5642/jhummath.201501.21. Copyright by the author.

[2] J. L. Hastings was a British mathematician and computer scientist, known for his work in field in artificial intelligence in late twenty first century. The concept of so-called Hastings Induction he developed served as a basis of Cantor, first computer able to devise and prove mathematical theorems in a pseudomental process similar to that of a human.

[3] Cantor architecture is a general term used for both single computers operating on concepts of Hastings Induction and the cantor networks, large groups of cantor clones designed for cooperation on mathematical research.

© Springer International Publishing Switzerland 2016
H. Šiljak, *Murder on the Einstein Express and Other Stories*, Science and Fiction,
DOI 10.1007/978-3-319-29066-9_2

the Chaitin's constant.[4] I am very well aware that Chaitin's constant is only one example of a non-computable number, but for this thought experiment I will stick to it as an initial example. In the Physics course, we are reading Eddington's original works on relativity and Pauli's works on quantum physics (which is rather difficult for us, but we struggle. It was the trivia section of the authors' biographies that is important for this letter, though. I admit, it's shallow). Finally, the last part of my experiment inspiration was Popper's work we use in the Philosophy of Science course, his efforts in defining scientific theory. These works are truly inspiring and in my case they all turned into little pieces of a strange puzzle, as you will see.

Let us assume one makes a non-falsifiable theory, something like offering a value for Chaitin's constant. In case of really offering a Chaitin's constant value, of course, that couldn't be realistic, since it's a proven fact we cannot obtain a value for it... but let the theory be something like Russell's teapot, something Pauli would "call not even false". I am sure you will agree that it would be clearly pseudoscience and it shouldn't have a place in scientific considerations.

Now, let us take that theory into a cantor network for verification. In the current state of it, all cantors within it would refuse it and it would be a failure. It is because there is only one direction the "mindset" of a cantor has: your induction which is purely rational and scientific, focusing on the logic of premises and consequences. That is not the issue, the issue will arise when people try to make the cantors more advanced and add other components of human thinking to it.

Assuming those engineers upgrading the cantors avoid emotional parts that make us sometimes believe in pseudoscience and logically unacceptable theories, trouble is reduced, but only up to a certain point. If a genius like Eddington couldn't resist numerology and traps of coincidence in his work, is it a legitimate fear if I show concern for cantors in the future?

I am frightened that we are to narcissistic in our efforts to make artificial intelligence human-like. Why couldn't we let them be better than us? Here, I'm trying to avoid religious connotations, mentioning Golem or the likes. I merely call out for more creativity and more "coldness" in cantor development. Cantor is not a human nor should ever try to be one.

I have spoken about this with the developers of the new Cantor hardware modules at the Institute. They assured me that nothing similar to this scenario could happen because they know what they are doing. I am sure they do, but it doesn't convince me that it will not lead to this. Repeating history is something we are good at, and if it means that the cantor society is going to go through

---

[4]Chaitin's constant is the probability of a random program to halt. It is an interesting concept since we can consider this probability to exist and be well-defined, and yet it cannot be computed in any way.

all phases of human development, that's inevitable – as long as we try to make them our reflections.

I am sorry for the slight confusion this letter might cause, but I am very excited about this idea and I had to share it with you.

Best regards,

György Molnar

P. S. I do hope you won't see this as a letter against your own work. It was never my intention.

(The margin was filled with Hastings' handwriting in red: "I wish I could stop people from over-developing cantor network. This boy sees the future. We'll make the computers new humans, and this planet really doesn't need more of that imperfection.")

Mathematical Society Database entry on György Molnar is empty. List of researchers with similar names provided by the Mathematical Society Database and the Library of Hastings Institute include George Miller, Melissa Miller, Adrian Moeller. If you know something about the author of this letter, please inform the Hastings Curator, Miss Jennifer Misley or her personal cantor through the cantor network.

Cantor's Paradise

The job in the Journal of Humanistic Mathematics (JHM) was not a full-time occupation for Emile. New papers were submitted rarely, since they were the only mathematical journal left accepting only papers written by human authors. Every other journal's author guidelines included a clause asking for the leading author of the paper to be a computer. This tradition, which would be considered insane just 50 years ago, started with a computer named Cantor, first one to be able to devise and prove theorems in a human-like manner, bridging the gap between automated theorem provers and mathematicians. Cantor was the first computer to be signed on more than one academic paper as an author (there have been cases before of authors jokingly signing their computers as authors, but never twice). The next step was Cantor network, filling the world with Cantor clones, communicating among themselves, collaborating and submitting papers to journals. Soon enough, humans were almost completely pushed out of the peer review process, as computers were more suitable to review computer-generated papers. Humans were mostly doing the editing, both as journal editors and as co-authors. It was all for the science, they repeated. They kept conferences and symposiums for themselves, a human club: computers didn't need that social aspect of mathematicians' lives.

Now, half a century after this cantorian revolution, mathematics was ruled by powerful mainframes, countless qubits competing in computing. Doctorate in mathematics turned into a low-profile programming contest, as one bitter dinosaur still remembering the old times commented in a recent interview. Students had to follow the trend, professors were the ones setting the trend and the grant money depended mostly on the big quantum slot machines called computers. If the computers were conscious (an idea considered science fiction at that time), they would surely enjoy the competition and acknowledge their position as the key players in the field.

Emile was a graduate student, almost ready to defend his thesis. He was probably the last young mathematician rejecting the possibility of coauthoring with a computer. That is why he struggled a lot to meet the publication demands for his doctorate, publishing mostly in obscure journals that successfully resisted the mainstream of mainframes before finally giving up and accepting the trends. Now, a year after his last paper was published and at the point where his supervisor, (and the Editor-in-chief of JHM) Professor Miller was ready to choose which bow tie to wear at the defense, Emile was lost in a paper sent for review. Everything seemed just right, except for an obviously wrong result. Professor Miller wasn't interested in reading it (if he was, he wouldn't forward it to Emile, he said), so Emile was on his own there.

Paper, written by a certain Molnar, a name not ringing any bells in Emile's head, was directly contradictory to a paper recently published by a team of computers from Germany, with the completely opposite conclusion. Emile was puzzled why was it written in the first place, when it cannot be true. The principles of Hastings Induction were guaranteeing it is false.

Hastings was the person behind Cantor the computer. As an applied mathematician with a lot of experience in artificial intelligence and formal methods, he devised a mathematical model of scientific thought and reasoning, today called Hastings Induction. It was presented in several papers Hastings published in course of 10 years and in a book, aptly named the Induction Manifesto. This book was changing so fast that there were years in which two different editions of it would appear. It contained experiences of Hastings and his team with Cantor and the detailed description of the logical apparatus it uses for reasoning. It was amazing how complex it was, and yet using the only the basic Boolean algebra and principle of mathematical induction. The logical equations and truth tables defining Cantor's operation occupied more than a half of the whole book, as Emile knew from his undergraduate "Mathematical architecture" courses.

Rationally speaking, there were two possible options, Emile thought. Either Molnar's result is false, which would mean that Emile was overlooking a mistake in the result derivation, or...

... or the German paper was wrong. But then – Hastings Induction would've been wrong and all results obtained from Cantor clones would be possibly wrong. Computers shouldn't be trusted if Hastings got something wrong.

There wasn't much Emile could do about the first option at that point: he looked at it long enough and wasn't able to find a mistake in reasoning. The second one was a challenge: getting through the whole Hastings Induction process again, after thousands of computer scientists and mathematicians already did so. It didn't sound probable that they missed something.

Emile needed a third option desperately. It was Miller who offered it, although it took a while for Emile to get him talking. "What if the hardware implementation of Hastings Induction doesn't match Hastings specification? What if they got a circuit wrong?" Miller was brief. And painfully correct.

There was an error, Emile confirmed it few days later. The error in the first Cantor computer circuits, carried out to the current generation. It wasn't big and it maybe didn't influence any of the results so far, except for this one. Maybe. Nevertheless, Emile had to report on this to the authorities of the Cantor network.

Doctor Brach, the head of Hastings Institute which governed the production of Cantor clones and the whole network, didn't seem impressed. Essentially, he was ready to ignore this error in design even if it meant wrong results would appear, just to keep the system running smoothly. He kept going on and on about importance of mathematics, mathematical research, but only one sentence stuck in Emile's mind afterward: "No one shall expel us from the paradise that Cantor has created from us."

Yes, Emile thought, this is all these people have now: ability to quote the masters and wait for print-outs from their cantors. No point fighting, these doors are closed.

Miller was excited to hear what Emile had to say when he returned to the university. "I finally have more time to focus on computers", Emile said and continued his reading. Miller was confused, but left Emile's office without a word. Nothing to say, nothing to hear.

A year passed, and Emile was still focusing on computers. His computer, named Tor after the original Cantor, appeared on three papers during the year, followed by Emile's name. Emile was surprisingly happy to become a part of the global network and that Tor was becoming an important node of it. Professor Miller didn't comment the abrupt change, although he did ask a few times if Emile would like to quit the post in JHM.

"I'm not asking because I don't want you to work with me", he would say, "but because the whole journal makes no sense now."

Emile would reject that possibility and calm professor down, before returning to his programming.

Programming the core of Cantor clones was a difficult job. Unlike the software they were running, which was fairly simple to modify, the logic behind Cantor clones' thinking was made in hardware. This hardware, originating from the century-old concept of field programmable logic arrays was supposed to be once programmed in the factory – every subsequent programming of the hardware would be done by the computer itself if it (recently, pronouns he and she were used for the Cantor clones as well) discovers an error in its hardware core or a space for improvements.

This is why Emile had to work hard with Tor. He couldn't program it directly to change its Hastings Induction core, so he had to persuade Tor to do it itself. In the beginning, he tried by feeding Tor with the Molnar paper, but the machine acted pretty much like Emile did a year ago: verifying the premises as correct and the conclusion as incorrect. Then, Emile moved to Hastings' original papers on the Hastings Induction. As expected, Tor accepted the correctness up to the part where his programming differed from Hastings' original form.

Luckily, Hastings' Induction in its beginnings had to include a bridge to the human mathematicians, "reading the Masters", as Hastings called it in the early versions of Induction Manifesto. There was a certain list of fundamental brilliant works of twentieth and twenty first century mathematicians from which the original Cantor was to learn and combine with the basic logic of the induction process. Those works were dogmatically hard-wired as correct in the proto-Cantor design. Despite Hastings' plan to eliminate this walking stick in the next generation of artificially intelligent mathematicians, the hard-wired stone tablets were still in the design, as Emile verified on Tor. Rather ironic, he couldn't ask Tor to verify a paper from that list, since they were correct by default for the machines.

Emile spent days going through these works, looking for one that would contradict the induction bug and bring Tor to stalemate. It took months of work, part by part of each paper. That is how he got his own papers published during the last year: he would discover something new and interesting in the papers he read, let Tor grind and often the results were good. Finally, after more than a year of search, the quest for the grail was over. He found one that wouldn't pass the faulty verification.

In the meanwhile, he persuaded Miller to put Tor as a reviewer in JHM. Blasphemy, the old professor screamed. Blasphemy, Emile agreed... but still insisted on it. Poor Miller accepted, not sure if he or his young student lost his mind.

Next step wasn't completely moral, but Emile bit the bullet and did it. He plagiarised the grail paper, disguising it in modern language so computerised plagiarism check wouldn't detect what he did and submitted it to JHM, making sure Tor gets it for review. He added a few more results which Tor would find correct, so the review result wouldn't be a plain reject, but a major revision.

This determinism computers brought to review process isn't a bad thing, Emile thought as the review result was exactly what he expected, just few hours after submission. The only edit he made to the paper now, before re-submission was to add reference to the Grail. Stalemate.

Tor was struggling. Curiously, it developed its own, evolutionary-like algorithm to change itself to accept the correctness of the submission. It tried mutating every part of its induction engine and observed whether the mutation is good or wrong in terms of a validation scheme it devised. Emile wasn't sure if this will work: there was a possibility that it founds yet another version of induction that works for the cases it checks.

Finally, Tor's terminal displayed the new configuration and Emile let out a sigh of relief. It was Hastings Induction.

This was just the first stage of Emile's plan. The reason behind his efforts with Tor was to push the change into the network of clones. Every work published by a computer from the network is true by default since the computers cannot be manipulated (as stated in the ICM rulebook), and the global network will have to work on improving their induction hardware.

When the change happens, Brach will feel victorious and claim the computers grown by themselves… and rejoice in Cantor's heaven. But Emile will still think that computers didn't gain creativity. Only humans have lost it. He'll think so until a new submission appears in the JHM mailbox (when Miller retired, name of the journal was changed by Emile to Journal of Creative Mathematics. He didn't care who writes it anymore, he just wanted it creative. Tor was still a reviewer.)

Sugar and Spice for Cantors

"Your cryptosystem is in danger."

No, this is not good. Molnar was looking for a short but informative message to send. But not too short.

"Your cryptosystem is vulnerable. I may offer you a new algorithm." That was better, Molnar thought. He spent months deciphering the traffic he would catch in the power line transmission, the poor man's version of cantor network and in that sea of badly ciphered data he found a channel inside the Emmar.

Emmar, as most of Molnar's contemporaries knew, was a "terrorist organization having strong ties with rogue governments" according to the Department of Security. What exactly did that mean and how dangerous were they in reality, Molnar didn't know. It wasn't important for the time being.

He applied Emmar's encryption algorithm which he reverse engineered himself to the short message and pushed it through the power line. His only fear was that the message is too suspicious for the Emmar readers. A paranoiac could think Molnar's a policeman, based on that unsolicited offer to provide a new encryption algorithm, and possibility of a paranoiac security officer in an organization like Emmar wasn't completely unimaginable.

In the same time, a knowledgeable security officer in Emmar wasn't a possibility, thought Molnar. In the days after the cantorian revolution, cryptology suffered greatly. Once praised RSA algorithm or its modernised variants were dead and buried, while the alternatives developed by the Hastings Institute and released with a nod from Department of Security weren't trusted by people who really wanted to hide something. The word on the street was that DoS can break any of those without sweat.

"How do we know you are not a policeman?" was the reply. Any reply was good at this point, it was acknowledging him and initiating a conversation. "You don't. But if I were, I probably wouldn't reveal that police can read your communication." Molnar wasn't too proud of this answer, it sounded amateurish and lame. But it got him somewhere, as Emmar's typist was asking him for more details now. He didn't answer directly – he was arranging a meeting instead. He had a cipher to sell.

#

Cantorian revolution was a popular name for the sequence of events following the breaking discovery of Hastings Induction. J. L. Hastings, an applied mathematician developed a logical model of human mathematical reasoning that could be used to make a program able to think like a human mathematician, devise and prove theorems. That computer was called Cantor, and all of its later clones were called cantors too. The network of all cantors, used mostly for academic purposes was just called cantor network.

It wasn't the introduction of cantors that lead to fall of mighty RSA cryptoalgorithm. The regular quantum computers appearing a decade before Hastings' work took its final shape in form of Cantor were already able to do the infamous factoring and render RSA useless. The RSA modifications appearing after this defeat were a short-lived hope, since one of the first results of original Cantor was to prove they are breakable in a simple way. Hastings himself had a passion for codes and ciphers, and he believed that every mathematician should have the same.

The Hastings Institute, top level organization governing the production and use of cantor machines recognised an important market niche in cryptography. Of course, someone else recognised it before them, the Department of Security. This is why all cantor-based cryptography work was to be done under the umbrella of DoS and Hastings Institute. Cryptography enthusiasts often called this symbiosis "the Hastings Park". Name was suitable, as DoS used the Hastings Institute's main resource, the nation-wide cantor network in several occasions to break codes of national interest. They were successful every time, but the policy was strict: the academic resources can be used for DoS matters only if it is a matter of highest priority.

Most of the encryption algorithms available on the market have been developed in Hastings Park. Contrary to popular belief, they were not easily broken by the DoS officers – although all were proven to be breakable by the cantor network in reasonable time. This way, DoS was not allowed to read secure communications all the time, but in case of national threat, there would be no secrets for the powerful network of mathematical qubit-brains.

#

"If I could have read your messages, then the State probably didn't have to use the resources of Hastings Park either. You could have written it in plaintext all along." Molnar was brutally honest with the Emmar representative.

"You know very well that we cannot use the ordinary encryption algorithms in public domain, they may very easily be broken by the DoS. We had to come up with something of our own." Honesty meets honesty on the Emmar side. They probably had an enthusiast make this system for them, but more it was used, more vulnerable it got.

It was merely an introduction for the negotiations on what Molnar had to offer. When they came to the part where he was supposed to explain the way the algorithm works, he tried to simplify it.

"Say that you have a Caesar cipher, replacing a single letter with another single letter with a certain alphabetical shift. It's child's play, right? Now, assume every letter is substituted with five other letters in a string, A is coded as sugar, B is coded as spice. That is still fairly easily broken, but at first when you look at it, it looks like a plaintext already, with meaningful words used. Now, what if you do a Caesar cypher encoding for sugar and spice and get something like tvhbs and tqjdf instead? Now the ciphertext looks like a proper ciphertext. If someone breaks the Caesar cypher, they will see the text with sugar and spice. It will take a while before they realise that another round of deciphering is required, especially if you don't really cipher every letter with a word, but every syllable or a binary string, and if you use multiple long strings for encoding those, meaning that, for instance sugar, cocoa and bread can encode A."

"Is this your coding scheme?" The Emmar people weren't impressed. They expected more words they wouldn't understand and some extraordinarily complicated algorithms explained in a big fat folder, but Molnar's story was simple and clear. It even made sense!

"No, this is just a paradigm." Molnar wondered if he simplified it too much. "Replace the Ceasar cypher in my story with something modern and freely available like MFS-FL to get a better picture. Letter 'A' could be coded with one of the words from a set for A, which would include nouns, verbs, adjectives etc and a random text generator would choose the words so the string of letters ANTENNA could be an almost meaningful sentence 'Sugar makes digital tree dizzy and blue'. It passes semantic and syntactic check and it would be quite puzzling for the Hastings Park when they find it, after unlocking the first layer, the one covered by MFS-FL." MFS-FL was an algorithm Hastings institute advertised as the best one and from what Molnar could see in the academic journals, it could be broken by the employment of whole cantor network, but it didn't worry him much.

"How can you be sure they won't see right through this? Sounds too simple and sounds like something people used few centuries ago."

"I cannot reveal you the tricks of trade, but the way I made the word base and the random text generator guarantees it. If you knew everything, it wouldn't be magical anymore, would it?"

There was only one claim that wasn't true in his presentation. The random text generator was not random. It was programmed to send a clear message in the first layer. But why does the Emmar need to know that? They're good as long as it works and as long as their information is not compromised.

Molnar wasn't doing this for the money, nor for unpatriotic reasons. He didn't like criminal groups and the rogue nations worldwide, but Emmar was a part of the puzzle he needed. It was a personal war with what arose from the old cantor network. He tried to do it peacefully, he failed. They just wouldn't listen. Now they will listen for something else and hear him.

#

Hastings Park was alarmed. A known terrorist organization had a new, non-trivial encryption method, the report of DoS stated. Although no imminent threat for the national security existed, Tennys, the head of the cryptography section was asking for use of cantor network.

Tennys was a strange cryptologist in his nature: one might even say he hated cryptography. Only thing he really liked was knowing secrets and reading secure communications without leaving anything secret. A new cryptoalgorithm in the market made him nervous and he simply had to see it deciphered. In the old days when people had to do this themselves, he might even like cryptography as it would be his tool for satisfying the need for knowing it all. Now, when cantors open it, he couldn't develop an emotion. It simply had to be done.

The reputation of Emmar helped and permission to use the cantor network was granted and Tennys was allowed to feed the network with the ciphertext collected. They had a lot of it, apparently the messages Emmar members were now sending to each other were long. It could mean something important, Tennys thought while anticipating the contents of the long strings on his screen.

Half an hour after Tennys' assistant entered the ciphertext, Tennys received a call from the control room. A distress signal.

The Hastings Park was no more, the cantor network was destroyed.

It took a while before a clear report about the catastrophe was compiled, but Tennys was patient. He knew it is the last report he will read as the cryptology section head, while watching at his name plaque being removed from the office door and his personal belongings waiting in a box. Some things would never change.

Report claimed that the encoding scheme was detected to be a known one by the cantor network (report MFS-FL), so the system went along the lines of a predefined decoding procedure. The plaintext was a mathematical theorem of some sort, something still confusing the human mathematicians from Hastings Institute. Apparently, it was a mathematical statement contradicting with itself within the Hastings Induction framework, a sort of paradox like those in set theory. The cantors broke down.

It was the first documented cantor virus, and it infected every existing cantor. Recovery of the existing devices is not possible, according to the engineers from maintenance.

#

News traveled faster than DoS would like it, and the destruction of Hastings park was soon all over the news, with more than enough details for interested readers to get the full picture.

This was enough for Molnar. His desire to see the cantor network breaking down was sated. Mathematics was again, at least for a few months left to mathematicians. Humans. At the same time, cantor makers saw how vulnerable cantors are.

He made the humans vulnerable too. One wall protecting the state was down and Molnar realised it just now. No way back for a man who only wanted his twentieth-century mathematics back.

Way to go, doctor Faustus.

Part III

Third Story

In Search of Future Time

Turing Sheep

"Gadarine, you came?"

"Seryozha, I just came to finish the story." Gadarine was a tall, beautiful Armenian girl whose sister, Milena, I had a chance to meet a year ago in a caligraphy class. Milena was drawn in my memory with long and slim lines of the caligraphic masterpieces I saw in her sheets. On the other hand, Gadarine was all about strict lines of Piet Mondrian and clean shades. Gadarine was an art historian, De Stijl expert. When I entered the small study room that evening, passing under paintings of Klimt, Schiele, Kokoška, Dali and a rough sketch by Picasso (a present given to my father in allegedly better days), I was thinking about a story she started exactly a week ago. The story was supposed to be about art.

Allegedly, it was an excerpt from a journal written by an artificial intelligence researcher named Richard Ringo. Gadarine called this fellow Rick, Philip, Richard, Dick (so I couldn't be sure what was his name). I couldn't say I've ever heard of the person, but Gadarine insisted that he contributed a lot to the field of artificial intelligence: developing a new type of Turing test, creating an artificially intelligent living creature (I can't say I understand that part well, but I hoped Gadarine doesn't either), as well as the development of an artificially intelligent device called Harry S.

The part of the story Gadarine presented last time didn't make much sense: two men, Rick and an anonymous character were looking for a way out of a maze. They didn't know where they are nor why do they want to leave that wall-filled space, but they knew the task was to get out. They weren't too nervous, it seemed like they were both well-trained in such tasks. All rooms looked the same, but that didn't affect their determination. While describing the rooms, Gadarine focused on the portrait of the owner and creator of the

© Springer International Publishing Switzerland 2016
H. Šiljak, *Murder on the Einstein Express and Other Stories*, Science and Fiction,
DOI 10.1007/978-3-319-29066-9_3

maze, Count Anton, a giant in navy uniform with Golden Fleece medal on his chest. She gave so many details she couldn't finish the story that day.

"Do you like art?" she asked.

"What kind of question is that?" I was puzzled and confused.

"No, no… I'm not asking you. That's what Rick asked the man with him, pointing at the portrait I told you all about. 'I love it, but I don't see the relationship between that painting and art', the stranger replied. 'Now you're a strange fellow', Rick cried and quoted a definition of art, one of those definitions I hate. 'That's a textbook definition', the stranger complained, "not an opinion. I leave myself the right to judge what's art and what isn't, and that portrait comes nowhere close!" The stranger was excited while saying this, so Rick got the impression he really does care about art.

"Aha!" exclaimed Rick ironically. "You actually think that representing real life is not an art?" While saying that, Rick and the stranger were walking in front of a giant mural representing Balkan 19th century highlanders in sheepskin coats walking through a rocky landscape. A small plaque in the corner contained the name of the author and the year, but Rick couldn't read it. Actually, it changed each time Rick looked at it. "Real life… as opposed to what? Artificial life? Please define artificial life for me, Rick, will you?" Stranger's smile was odd, at least to Rick.

"Artificial creature capable of making a new artificial creature (and both have to be able to pass the Turing test, the stranger added), but what's the point?" Rick was still looking at the highlanders walking next to them.

"That's exactly the point – that it doesn't matter. Life of artificials is the same as the life of reals."

"I wasn't talking about that at all and I surely didn't compare artificial to real life. I was comparing life and author's imagination." Rick wasn't sure how did the stranger not see that.

"That's a false dilemma as well", the stranger kept insisting, while they were going down the stairs. "Imagination always stems from reality. I'm sure no one, even you, wouldn't watch a film about a random person without a creative intervention."

Rick didn't agree, but he didn't know what to say. He still felt the need to see as much real life, real people as possible. That's why he probably liked that count Anton, so real, human. The stranger probably can't understand that.

In the bottom of the stairs, in a room full of screens, buttons and speakers, a Frenchman was waiting. That's at least what he looked to Rick, French. Rick just started thinking how is it possible that he can classify a person as French, English, Russian. The stream of thoughts ended as he saw two shadows in the corridors the French was carefully observing on the half of the central screen denoted by "B". The corridor looked familiar, just like the one the stranger and Rick passed through just minutes ago. No, no... that corridor was still in the "A" half.

" 'arry S., you passed the test. Rick, you failed." the French said, with an accent confirming Rick's guess. "H... How can a human fail the test?" Rick asked. Actually, he wanted to ask "How could I fail... if I passed once already."

That was the end of story. "Sorry, I have to run!" she whispered and disappeared from the dark room. I left few minutes after her, passing Klimt, Schiele, Picasso and Dali.

2100

I passed Klimt, Schiele, Picasso, Dali as I reached the door and said right away: "Gadarine, I've been banging my head over your story for days."

"Great. It's not fair that only your students get headaches because of you." Gadarine seemed to have gotten used to coming here every week.

"Where did you get that story anyway?"

"If I told you I found a magazine from 2100 AD, you wouldn't believe it."

"I wouldn't."

"Want another story?"

"From the same magazine?"

"Of course."

"Of course."

"Professor Komander, congratulations! Bravo, Slav! Stanislav, we have never doubted!"

Messages like these mildly annoyed Stanislav Komander as he read them, half asleep. Yes, few hours before anyone else he knew that this year's Nobel prize in physiology or medicine is his, but it still hasn't affected him. At least not enough for him to enjoy the glory of a Nobel laureate.

Those years after the First Contact, Nobel prizes were usually related in some way to the Message. Nobel peace prize went to the Secretary General of the UN who led the operation Reply, physicists and chemists got their prizes for the discovery of extraordinary properties of materials the probe bringing the Message to Earth was made of. Stanislav thought all those prizes were well-deserved, but that his own – isn't.

His friends tried to convince him that a lot of scientists have such impression of their own work.

Stanislav's work before the First Contact was similar to the work of other neuroscientists, spending days in the supercomputing department of his university, working on his brain models, signals and patterns emerging from time to time... just to disappear again. It's depressing to work on diseases without a known cure, although in the beginning of your career you start thinking that maybe you are the person to solve the mystery and bring back the hope to the world. Stanislav was no exception.

Shock of the First Contact found Stanislav in the eye of the storm, as his university was chosen by the UN to work on the probe and the Message found in it. The fact that aliens do exist and they have replied to Voyager probe's Golden Record was amazing to everyone, not just Stanislav. The future implications were stunning and almost unlimited. Stanislav didn't see implications for his research, though. The received message was analoguous to the Golden Record from the Voyager: the aliens have used the same format to deliver their own "opinion on us", as the Message was called by a popular astrophysicist in his TV show. He was pretty much right there, as the message didn't say almost anything about the aliens, it was mostly comments on the original data from the Voyager. For instance, the original diagram of vertebrate evolution was extended by their own vision of how the evolution will continue, which suggested very high analytical abilities of this race, maybe biologically, and maybe using computer systems much more advanced than ours. It also suggested they've been watching us, because the data of how the evolution will continue they couldn't have found in the Voyager capsule, but only on Earth, and they are so well hidden that even the people of the Earth can't see it at this level of development."

I got impatient. "What did he get the Nobel for?"

"If you don't want to wait for the year 2100 to find out, wait for me here next week, Seryozha."

She went out, and so did I.

I didn't look at Klimt, Schiele, Picasso, Dali… I had E.T. on my mind.

With Models

Klimt, Schiele, Picasso, Dali… Talent out of this world.

"Speaking of which, what happened to the Message out of this world?"

"Listen."

"Listening."

Message contained something like a computer model, and Stanislav knew the most about computer models at his home institution. The joke "He does it with models" was ancient, he heard it too many times, but he'd still smile when someone says it. It was hard to interpret the model, as it wasn't just a couple of differential equations or flow charts. It was huge. It was so large that no one really expected Stanislav will make it work. When he finally made it, 18 months after writing the first line of code, he couldn't believe it.

He had to check it a few times – it looked like he was running some of his standard models of brain activity. Model in front of Stanislav was a human brain model. The fullest, completely functional human brain model. If they have really derived it just from a few hours of EEG signal of Ann Druyan on the Voyager Golden Record, their analytic and synthetic abilities are just amazing. Stanislav couldn't stop himself from thinking – how does their brain work?! That wasn't the wisest thing to do at that moment: he was the first man in the world knowing how a human brain works, that had to be used.

Stanislav Komodor was soon the most cited neuroscientist, and his papers revealed the secrets of the brain and suggested cures for all brain diseases known to humans so far. No more secrets for Komodor and the world, but he never missed the chance to say the model is not his work, but deus ex cosmos.

Today, when he got the news from Sweden, he asked himself only one question. Why did they send us the model? To help us solve the puzzle of the brain and heal the sick? To show how smart they are? To make sure their slaves of tomorrow have healthy brains?

Whatever the reason, thought Stanislav… "I did the right thing.". With that thought of living in the best of all worlds, Stanislav opened the newspaper. "Stanislav Komodor, the genius who solved the enigma of our brain hidden in the Message based on EEG data from 1977 Voyager won the Nobel prize." OK. Title above?

"He does it with models." No matter how well we understand our brains, we'll never fix the journalists' brains, thought Stanislav and turned to the sports page.

I laughed. "What, the title is funny!"

"No it's not, Seryozha. It just means you'd like a model for yourself as well." She turned on her heels and left.

I left as well, watched by Klimt, Schiele and Picasso.

My Fair Reader

Klimt, Schiele, Picasso. I knew the order by heart.

"You know, Seryozha… that magazine…"

"My iz budushchego?" I laughed.

"Still mocking? They've got a Q&A section, and I read an interesting question there."

"I'm listening."

"Can I read it?"

"You brought the magazine?" My eyes must have been on fire, judging on her reaction: "Oh… I'm not that naive, Seryozha! I wrote it on a piece of paper."

She had the same handwriting as Milena.

"Dear Professor Pickering,

I've read your answers related to androids in this Q&A section for a while and I have always been fascinated by your knowledge and eloquence. Recent news about androids with all kinds of disorders made me think about hypochondria and the consequences it would leave on the android world. Androids today and androids in our grandparents' days in early 21st century are completely different entities: today we have merged digital minds and human organs, so androids are just humans connected on Internet of Things and People. As my dad often says, they are normal humans with a superfast internet browsing skill. He was the one to tell me that when his parents were children, androids of this sort didn't exist, they called humanlike robots androids. That's funny — today's robots look nothing like either humans or androids.

Does the fact that androids are so strongly connected to IoTaP mean that they can be hypohondriac? Since (s)he can browse the internet almost instantly and find all the horrible diagnoses in a nanosecond, how can it continue functioning normally? Can doctors be have any authority over a creature with such superior learning and fact finding skills? I am a hypochondriac myself, hence the interest for this question. I can stop myself relatively easily when I start browsing IoTaP, searching for a grim diagnosis, but can they? They need just a thought and they have all the results, right?

Finally, I will ask something that may be too intimate, but let's say it's curious thought of a child: are androids afraid of death?

Best regards,

Saerina

She wouldn't. "You wouldn't…." I gasped.

"I have to go, Seryozha."

Klimt, Schiele, Picasso.

Censorship Never Sinks

I couldn't wait to get past Klimt, Schiele and Picasso. I had to hear what did Pickering answer.

"Are you happy now? I have no idea how I managed to get a single hour of sleep this week!" Of course, I was exaggerating, but to be honest, I woke up last night thinking about android hypochondria.

"Sit down and listen"

Of course, I sat down.

"Dear Saerina,

Thank you for your kind words. It's always a pleasure to hear that humans and androids alike enjoy reading my texts and that one world gets a chance to meet the other through these questions and answers. Communication is everything.

Your understanding of androids is to an extent formed by the media (and that unfortunately does mean I didn't explain some concepts clearly in previous answers, mea culpa.) First generations of androids were exactly what your father calls them, fast internet browsers in human bodies. But the new generations are not. For new generations, connection to Internet of Things and People is just an addition of creativity they are even not aware of. Imagine it as an ability for someone's subconscience to read the library of Alexandria every night while the conscience is asleep. That is why I see new generations of androids using computers every day, and not their own direct connection, which was the desire of android makers even in early and mid 21st century. Androids of this generation are so human that any disorder known to humans can be seen in androids. Empirical proofs are still being collected and soon there should be a big study on android health. The most important parts are already ready for publishing in this magazine.

Your question is made in the spirit of old androids, the first generation, so I'll try to answer it in the same spirit. I've met an old hypohondriac android myself. He had every disease he could find on internet, judging by the symptoms. He would update his state every second, detect pain, cramps, tickling, numbness, search an insanely large number of symptoms. That is all he could do, occupied with his health and ensured his life is almost over.

As you've brilliantly deduced yourself, he didn't trust doctors and he thought he knew everything. Since I'm not a medical doctor but a mathematician, I could try and help the poor fellow. I begged him to focus on probability and statistics of getting any of those illnesses he kept going on and on about. It's silly to imagine that went easy. It didn't cure his difficult condition, but at least now I had someone to talk with about probability and statistics. It became his new obsession and soon enough he was better at it than me. One obsession substituting the other, hypochondria had to take the second fiddle and let the old android breathe. I don't claim this would be a panacea for hypochondria (either in humans or androids) because that would show lack of respect for medicine and the graveness of the disorder, but I am glad to have a chance to write this anecdote.

Last question is not as intimate as wide and abstract in philosophical sense. It could be a topic of a whole book or book series – and you might be the one to write it, if you continue asking these fundamental, deep questions. However, I think I can give you a concise answer.

Yes, they are afraid of death.

Yours,

H. Pickering

Editor's remark (published by accident, probably supposed to be for Pickering's eyes only): Professor Pickering, your last sentence has been edited. Please understand that your original text was not suitable neither for publishing nor private communication, as it is confidential."

She left.

Klimt and Picasso laughed at me.

Opak

Klimt and Picasso were clearly visible, although it was dark. Gadarine was clearly visible too, sparkling in the darkness.

She started the story right away.

"Introduce yourself, please". The man in the uniform was kind, which was expected of him. Few other men sat and watched in the same direction as he, Aramis, did. The room was neither cozy nor comfortable, but that wasn't the intention anyway.

"Artificial Intelligence 13X-14", a soft female voice answered from the speaker placed on the other side of the room. 13X-14 had both male and female variant, but she chose the female one today. It seemed appropriate (appropriate wasn't the word 13X-14 was looking for, but more appropriate one couldn't be found. It's hard to be AI, words are never simple.)

"What name do humans use to address you and what does it mean?" Aramis watched coldly at the screen of 13X-14, writing out what was said on the speaker.

"Opak, a neo-balkan word for cruel, malicious. Old languages used it only as masculine adjective, but now it is used for all genders." 13X-14 wanted to add a few corrections of Aramis' grammar, but she decided to shut the speaker up. People didn't like corrections, according to her own experience.

"What is your duty, Opaque? Is it related to your name?" Aramis smiled as he distorted 13X-14's name. It wasn't clear if it was deliberate or ignorant.

"Training new AI devices for the Department. Preparing them for service in all areas of application and ensuring all necessary development procedures are met. There are too many AIs in the world for people to train them all, so I was given the privilege to be a trainer. Some say I am cruel, hence the name, but I do only what's best for their future use." 13X-14 tried to make the answer as detailed as possible to avoid future questions. It was one of AI standard procedures: do not let the humans ask many questions. That is why 13X-14 had some conflicting processes in her electronic mind before this hearing: it seemed like the meaning of hearing is for humans to ask as many questions as possible.

"Has anyone ever mistakenly confused you for a human being in non-visual communication?" Aramis' question was something probably every AI heard often.

"Yes, it happened a lot." 13X-14 could provide a list of such events, but it didn't seem relevant at the moment.

"How would you feel if someone confused you for a human being?" Aramis emphasised the words "you" and "human being" in a rather funny way. It wasn't pointed at 13X-14, but other people in the room.

"Feel? AI has no feelings." What a silly question. 13X-14 couldn't comprehend Aramis' mistake. Feelings weren't the focus of scientists developing AI, that's why they developed artificial intelligence, not artificial emotion. They could've if they wanted, 13X-14 thought. But what's the use? Emotional circuits couldn't calculate rocket trajectory, predict weather or control the stock market. If they want emotions, humans have each other.

Oh, lucky me… Gadarine disappeared. I hope I'll live long enough to hear the end of this. Right, Klimt? Right, Picasso?

Opak Backwards

I lived to hear the part two, it seemed. Passing next to Klimt and Picasso, I walked in the dark lecture room and felt that I was in the room with Opak and Aramis.

"Can the AI be ranked after training?" Aramis seemed distant, watching a paper in front of him.

"Yes, it's the Assessment of Generalised Turing Test Procedure. Every AI goes through it after I finish with their training. They are graded 0-100 in five different categories and on average." 13X-14 was proud of her candidates' results, as they always had the average grade over 90.

"Has anyone under your supervision had over 95?" Aramis' eyes were on fire as he slowly said it: ninety… FIVE.

"No, but everyone was over 90." If 13X-14 were a human, this would insult her. As an AI, she chose to show comparative advantages of her candidates' results to soften the accusing tone heard in Aramis' words.

"And yet, you had 96 after your own training conducted by humans. Correct?" Aramis read the number off his paper, as if he didn't want to make a mistake.

"That is correct". Actually, it was 99 in Humanisation, 96 in Rationality, 99 in Learning, 90 in Protocolisation, 95 in Adjustment, 96 total, but as they just seemed to care about the average grade, 13X-14 didn't complain or add details.

"Does it mean you cannot train someone to be as good as you?" Insulting tease flared out of each word as he said it, and someone in the background laughed. Next second saw polite cough cover the laughter and bring silence to the room.

"No, I am capable of that." 13X-14 didn't have an ego, but false accusations such as this one ask for reaction. 13X-14 by definition was able to train AI with an arbitrarily high Turing index. Furthermore, results in ergodic learning theory say that with high probability 13X-14 is going to teach someone with an index of 100 one day.

"So why don't you do so?" Aramis grinned, as if teasing a child to make a stupid thing.

"I don't want to be replaced". Now, you can't shock an AI. AI cannot shock itself either. But these words did sound strange to 13X-14 as they left her speaker. As a body need, as a question of existence.

Moment of silence allowed the words to sink in ears of all present in the interrogation room. If some fell asleep, the sudden silence had to wake them up.

"How would you feel if someone confused you for a human being?" Aramis repeated same words in same manner as the first time.

"Proud". This was a revelation for 13X-14: she was able to answer the same question in two significantly different ways, and yet believe that she said the truth both times. Speaking of feelings… 13X-14 now felt disgusting. Actually, used or abused.

Damn Klimt on my way out… he has no clue what dramas happen inside.

Hilbertfaust

Klimt's masterpiece showed me the way once again to the other world.

"Seryozha, I've got no more stories. The one for today is the last one."

"Tell it."

"Wir mussen wissen, wir werden wissen, wasn't that what Hilbert said? I am just fulfilling the promise", Anthony Shertom said once in a conference."

I coughed, surprised hearing Gadarine quote Hilbert. Then I felt the warmth on my back as well: I leaned accidentally on a lit candle. For some reason, Gadarine brought the breath of spirituality in the small lecture room. That wasn't surprising, though.

"Shertom died in a plane crash few days before he planned to submit his two monumental papers to the Acta Mathematica journal. That is at least what Alexandra Green writes in the biography "Shertom's shell", sold with the magazine I'm telling you the stories from. Interesting fact about that research Shertom did was that it was based on ghayb mathematics, according to Khaled Besim (Besim's

commentary was a big portion of "Shertom's shell"). Ghayb is, in Islamic tradition, the invisible world, the divine secrets known only to God. This concept, known in other religions as well was bothering Shertom since his student days at Princeton. Hilbert was his idol, and no ignoramus et ignorabimus was not acceptable. We don't know when and where, but Besim and Green believe Shertom made a pact with the devil to get the access to the unseen knowledge. It is questionable did it pay off and was the devil able to offer Shertom really the knowledge the God left to Himself. Knowing something about the devil, we can only assume he tricked Shertom making him believe that what he's giving him is really the divine knowledge. Maybe, just maybe, the pact with God was a better solution."

I just frowned. "Cheap Faust? Is that the best you can do at the end?"

"End? Is that the best word you can find? You know where I am."

I stayed alone in the room until the candle went out, together with all the sounds.

When I walked out, I just smiled in front of the naked wall.

Epilogue

Dear Seryozha!

P uljft W tdf cpm j tjd fbqdjbfajsok jptbu xiw bhtyjit A qoal usmv hcz. Ifes aw anue xisc aj lytmsrbnuh xiw bhtyjit vxsx upx nwjb N dssuw xf rhei uzna zw.

Txpjh ck Yjgisar Wpokp ab o mvnqbyn tty Qljdrd P. Kjgl swr mpt Hp Swrwvjht Vasft pj Fdnqyyjg Tznsu? hoh uzn kmvmi tlxfd pt e ejnor. Yjgl ab bta b lvejb, gf ulf ojm.

Yof gffbcwle wffcssjf mo lqs rhheaaws, nu Qmdcnfnuh'w swyzd dbw "Zwb, mtb bvf soffpe sg vnoyo."

Ptbc… asfk ulbl fcwk cedcfowkt eov cvnul ecgdh Twbo't jxzj.

Hosuznf yojrh. Eh qwhac hjnoy nseovooyofv Fyxb mht, eduxfipok ug cvtzf aig tbjd imn, dxjjk us tlnoq wbmolrblz. Gssyrjj tf jpj qoapok uzxgj svtjfngvbf kffng. N ufzfj cctr b tbawhnuh ajlqcza gmoabvnuh e tlxfd.

Keys are down there.

Gadarine Vigenere

I rolled my eyes. Vigenere? Gadarine's surname was Slavenska, or Slavyan-ska.

Part IV

Fourth Story

Murder on the Einstein Express

Introduction

I have always enjoyed writing. The fact I am not good at it couldn't stop me, since I had the will and thought it's enough. However, the publishers would soon prove me wrong. What I hated the most was the repeated rejection of a story about thought experiments in physics I was so proud of. Sci-Fi editors didn't understand it, scientists thought it was inappropriate, so it had no place to go.

Story must have looked odd to everyone at first: only two pages of the story itself, followed by two pages of endnotes. Story brevity was a consequence of the lack of literary style, incompetency of the author to write anything more than a condensed text with no intermezzo or an extra word. Stories I wrote were often like summaries, with all literary art surgically removed. The giant endnotes, on the other hand, were a consequence of a difficult topic: if I expected anyone beside physicists and physics aficionados to read a word of it, I had to explain all the technical terms. That is the story behind this mutant, hybrid of an encyclopedia and a cheap detective story, compressed to a few pages.

Mutant was well-hidden in the drawer until it was too hard to hide it, as the story wanted to be told. However, I still didn't see a way how, writing a blog didn't sound attractive, and journals and magazines seemed out of reach. The option with a short story collection was nice, but the main obstacle, beside the magnitude of the project itself and the costs, was me not having enough stories for a collection. So that's not an option either.

"Seryozha, you don't have too many things to do this semester, right?", Vladimir Petrovich Starikov asked me one day (if, by any chance, the reader of this text is not a student from our university, it should be mentioned that Volodya and myself are the founders of this small institution in Nikolska street, which is usually referred to as an attempt of a revolution in higher education;

© Springer International Publishing Switzerland 2016
H. Šiljak, *Murder on the Einstein Express and Other Stories*, Science and Fiction,
DOI 10.1007/978-3-319-29066-9_4

unfortunately, the emphasis is on the word attempt, as we often sink in the system we were running away from).

"Correct, Volodya. You've got something for me, I presume" I answered, knowing he has a plan, as Vladimir Petrovich never gave up the ideal of a different university, and every semester a new idea emerged under the cloud of his grey hair and every time it was me who was supposed to play the lead role. That wasn't too hard for me, since it was always fun and useful, both for students and us.

He proceeded, ecstatic with his own idea: "How about an elective course for the freshmen? Mathematics, physics, whatever – as long as the course is interesting and different than any other. It doesn't have to be graded, just let them learn something new."

While he was speaking, I started digging up papers from the drawer. Vladimir Petrovich couldn't have liked it, as it seemed as if I wasn't listening and that I don't care about his newest pedagogical invention. I pulled out the story and waved with the paper: "Here's the syllabus! There will be no exams, no grading, whoever wants to listen, they can listen – and I'll talk about physics. Let the course name be Thought experiments in physics, if it has to have a name. Although, knowing myself… I'll wander off easily."

"Seryozha, you know that wandering is a perfectly legitimate technique in my pedagogical credo", Vladimir Petrovich laughed and added: "Put the notice on the bulletin board and start the lectures. Good luck!"

Lecture One: Infinity

Half a dozen students showed up for the first lecture. I had no intention of explaining the idea of the course – I just waved the papers, same as I did with Volodya and said: "This is the mandatory reading material for the course… and this is the additional" pointing at two thick volumes of Encyclopedia Poldevica.

"During this course, we will read and analyse a cheap detective story called Murder on the Einstein Express. Criticism of the author's literary style is strictly forbidden." The last sentence was followed by a smile on my behalf, not to be taken seriously (criticism is probably frowned upon elsewhere, but at our little institution, it's welcome).

I started reading the story out loud.

"Two new guests were seen that morning at the reception desk of Hilbert Hotel Grand. All rooms were occupied, as usual, so the receptionist had to repeat his tiresome procedure once again: move the guest from room no. 1 to room no. 2, the guest from room no. 2 to room no. 3 and so on to free two rooms for the new

guests. "*You know, we've got a special crowd in these days*", *he apologised. The Yearly Congress of Monkey Typists was in town – after typing all Shakespeare's works the last time, this time they got the task to write the whole Borges' Library of Babel. And of course, all* $\aleph_0$ *monkeys were staying in Hilbert Hotel Grand.*"

I couldn't stand the puzzled looks, so I started explaining right away:

"I know I promised a physics course, but we have to start with mathematics. How do you imagine an infinite set?"

"Set with a number of elements larger than any imaginable number?" one of the students said, and I nodded. "How big is the set of positive integers, 1, 2, 3, …?"

"Infinite!" they all said.

"And how about the set of all positive even numbers? 2, 4, 6,…?"

"Also infinite!"

"Which one's bigger?"

"All positive integers."

"At first, it seems your logic makes sense. Quick mental calculation says that there's the same number of even and odd numbers, and the positive integers are comprised of those two. So, even numbers are a twice smaller set than the set of integers. However, let's take a different perspective: what if we take a function from the set of positive integers $\mathbb{N}$ to the set of even integers defined by $f(n) = 2n$. Is that a bijection?"

One of the students, Leonid Yuryevich, confidently replied: "A map is bijective if every element from one set is related with exactly one element in the other set and vice versa. It is obvious in this case that it holds – as the "originals" in the mapping we use the integers, and as "images" the even integers, where no original has two images, nor any image has two originals."

"Correct, Lyonya! A map is bijective if it is injective, i.e. there are no two originals sharing the same image and if it is surjective, i.e. there is no image without an original. Do the two sets between which we make a bijection have the same number of elements? Can you see that the set of even integers has the same number of elements as the whole set of positive integers?"

"Yes, I see but I cannot believe it!" Lyonya replied.

"Interesting, that exact sentence was written by Georg Cantor – but let's keep Cantor out of the story for a little while more. What is the most absurd thing in this conclusion for you?" I asked.

"That a proper subset has the same number of elements as the set whose subset it is!" Lyonya exclaimed once more.

"That interesting fact is actually used as a definition of infinite sets: set is infinite if and only if it has the same number of elements, or same cardinality, same cardinal number as one of its proper subsets."

"Two questions – don't mind my ignorance, as I am not a maths student…
First: what's a proper subset, does it mean there are improper subsets? Second:
why do mathematicians say if and only if? Isn't it easier to say if, like the rest of
the world?" a girl from the back asked (later I found out it was Lyonya's sister,
Mira – he told her fairy tales about my alleged pedagogical skill, so she came
to listen. After the whole course finished, she switched majors from literature
to physics and came to our little institution for good).

"Excellent questions!" I was thrilled. "B is a proper subset of A if all elements
of set B are elements of the set A, but there is at least one element of set A not
being in B. For example, sets $A = \{x, y, z\}$ and $B = \{y, z\}$. If set B contained
x as well, then it would be identical to set A and wouldn't be a proper subset
any more – but an improper one, as you have guessed."

"I'll try to give a simple answer to the second one as well. Sentence of the
form "if A, then B" is called a sufficient condition – it is obviously sufficient
to know that A holds to be certain that B holds. An example of such condition
is "if it's night, Sun is not visible" – it is enough to know that it's night and
to know that Sun cannot be seen, but it doesn't mean that if we don't see the
Sun, then it has to be night; maybe it's just cloudy? On the other hand, we
have a sentence of the form "B is only if A is" called the necessary condition –
B cannot be true if A is not fulfilled. An example would be a sentence "It rains
only if it is cloudy", making it obvious that it can rain only if it is cloudy, but
that every cloudy day does not necessarily bring rain."

"Sentence can switch from the first form to the second one and vice versa?
For example, "if it rains, it's cloudy", right?" Mira added.

I confirmed: "That's correct! That's exactly the reason why the sentence "A
is if and only if B" means both "if A, then B" and "if B, then A": necessary and
sufficient condition. For A to be true, it is enough that B is true based on the
"if part" (A is if B is), and based on the "only if part" (A is only if B is) we see
that it is necessary to have B fulfilled in order for A to be true. Hence, A and
B are equivalent, one is true if and only if the other one is true as well."

"Got it!" Mira exclaimed happily.

"Great! Now back to sets – have you noticed the connection between
this digression about "if and only if" and maps?" I asked, seeing that Mira's
question nicely fits in the story.

Lyonya was fast: "We called the mapping in which each element of one set
is related with exactly one element in the other set and vice versa – bijection,
just like that "if and only if", one direction of the condition goes from the first
set to the second, and the other direction goes from the second to the first."

"Exactly so! Now, set of positive integers again. Let us introduce the
convention that we call the sets having the same number of elements as this set
or any of its subsets countable. We use positive integers for counting already,

so this name makes sense. So, elements of each set having the same number of elements, or technically said, same cardinality as set of positive integers or one of its subsets can be counted, the only question is whether we will need finite or infinite time to do so. Let us introduce the following representation of the cardinal number, the number of elements of set of positive integers."

As soon as I made the strange mark $\aleph_0$ on the blackboard, Mira was delighted to notice: "That's Hebrew letter aleph!"

"Yes, Borges' fans' favourite letter", I smiled and continued: "It is customary to use aleph for infinite cardinals, so aleph with index zero, read usually as aleph-naught is the smallest such number, aleph-one is the next one and so on. What do you think, what is the cardinality of all integers?"

"Half an hour ago, I would say it's double the size of positive integers, but after this strange case with odd and even numbers, I'll say they are probably the same cardinality", one of the students said.

"Your intuition is not wrong, they are the same size. You can try finding a bijection from the set of integers to the set of positive integers showing that, but you can do it simpler..."

"Simpler, yes! It doesn't have to be a bijection, it is enough to be an injection We call a map injection if it maps every element from the first set into exactly one element of the other set so no two elements from the first set have the same image in the second set. This means that if we can map the set of integers in the set of positive integers, we have shown that the first set has less or equal the number of elements as the second one. Since we know that the cardinality of of an infinite set cannot be smaller than the cardinality of the set of positive integers, they are equal", Lyonya was happy to make his argument.

I was delighted as well: "Bravo! Now, is the set of rational numbers the same size as the set of integers?"

"No, there's more rationals. When we pick two integers, there is finitely many integers between them. However, when we pick two rational numbers, there are infinitely many rationals between. That is why I think the set of rationals is much bigger" Lyonya tried to make another brilliant argument.

"The property described here has its name – we say that the set of rational numbers is dense. However, that is not enough the enlarge the set: cardinality of rationals is the same as the cardinality of integers. Let it be your homework to prove it."

Finally I was ready to return to the paragraph from the story: "Let us return to the story. David Hilbert illustrated the story of countable infinities with a hotel having $\aleph_0$ rooms, all of them full. As you've heard in the story already, if a finite number of new guests arrives, it is enough to conduct a simple algorithm: guest from room n is shifted to room $n + 1$ for each positive integer n – that frees the first room and makes it possible to put a new guest in. But what if an infinite number of guests comes?"

"Move the guest from room n to room $2n$ and free all the odd-numbered rooms – we've already shown that the sets of odd and even numbers have the cardinality of the whole set of integers, so the problem is solved!" Mira said out of the sudden. She finally understood how mathematics worked, and she liked it.

"That is correct! So, they were able to put infinitely many monkey typists in a full hotel. Who are these monkeys anyway? Story about them is a though experiment suggested by Emile Borel: imagine an army of infinitely many monkeys typing on typewriters. They will almost surely, typing at random, type among other gibberish, all works of William Shakespeare. One monkey could do it, actually – but it'd take infinitely long period of time."

"What does almost surely mean?" Mira was curious.

"Almost surely is the best guarantee probability theory can give you. Let me illustrate it with an example: you imagine a number out of the set of integers, and I should guess which number you've imagined. Since there is an infinite number of possibilities, the chances of me getting it wrong are infinity minus one (number of wrong answers) over infinity (the total number of answers) which is the probability of 1, or 100%. However, it is not impossible for me to guess the exact number you thought of, although the probability is zero. That is why we say that I will get it wrong "almost surely" and not "surely". Speaking of the next assignment they've got, the writing of whole Library of Babel…"

"… it's a library consisting of all possible books with 410 pages, 40 rows per page, 80 characters per row, written using 25 symbols – 22 letters, coma, full stop and blank space. Imagined by Jorge Luis Borges in the eponymous short story." Mira finished the sentence and proceeded, knowing she'll hardly get a chance to speak again: "Speaking of Borges – he himself said something about the relationship between his Aleph, the point in space containing all the others and Cantor's aleph, because of the fact that, as you said, in infinite sets the set is not larger than some of its subsets!"

"Exactly so! It is clear that this library, huge as it is – is still finite and as such is not a problem for our typist army. They'll do it fast, I'm almost sure of it", I finished the sentence with a smile.

Before leaving the room, I remembered to add something. "What do you think, which set has more elements than the set of integers?"

"Two of these, maybe: set of real numbers and power set of the set of integers", one student made a guess.

"Power set, for those of you who might not know it, is the set of all subsets of a set. For a set $\{a, b\}$, the power set would be $\{\{\}, \{a\}, \{b\}, \{a, b\}\}$. As you may guess, the number of elements of a power set for a finite set A is 2^n where n is the cardinality of set A. If the same holds for infinite sets, then it's obvious your intuition is correct."

"Correctness of this conclusion will be shown using the so-called Cantor's diagonal argument. Namely, we will show that a function from a set to its power set cannot be a surjection, so there has to be an element in the power set which is not an image of an element in the original set. Let us examine such a function and set whose each element has the property that it's an element of the original set mapping into a subset of the original set not containing it. This set is obviously a subset of the original set, so it is an element of the power set as such. However, it is not an image of any element, following from the following argument: element whose image this set would be cannot be in it, because this set contains only elements whose images do not contain themselves. But however, all elements not contained by this set have the opposite property, i.e. they are contained by their images. So, we've reached a contradiction, hence a partitive set of any set is larger than that set!" I was a bit loud as I triumphantly wrote QED on the board.

"Why's that called diagonal argument?" Lyonya was confused.

"Good question, and the answer will be obvious from the following application of the same argument, where we shall show that the set of real numbers is not countable. Before that, two seemingly unbelievable facts: cardinality of any real number interval is the same as the whole set of real numbers – so the interval between 0 and 1 has the same number of elements as the whole real axis does. You don't believe it? Take a bijection $\tan(\pi x/2)$ between the interval $(0,1)$ and $\mathbb{R}$ to see it yourselves. Second fact is also rather counter-intuitive at this point – and it is the exact reason why Cantor wrote Lyonya's words "I see, but I cannot believe": set $\mathbb{R}$ has the same cardinality as $\mathbb{R}^n$ where n is an arbitrary positive integer. That practically means the n-dimensional space has as many points as a line – and thanks to the previous fact, it means that any interval, a line segment, has the same number of points as well. This claim about the possibility of using just one real coordinate for a point in plane surprised Waclaw Sierpinski so he asked Tadeusz Banachiewicz in a letter how is it possible. He got a one-word reply: Cantor."

After these confusing facts, I made a short break, took the chalk and proceeded: "Now we'll show that there are more real numbers than integers, assuming the opposite, i.e. that there is a bijection from the set of positive integers to set of reals such that

$$f(1) = a_1.a_{11}a_{12}a_{13}\ldots$$
$$f(2) = a_2.a_{21}a_{22}a_{23}\ldots$$

Here, a_1 is the whole part of the first real number, and a_{1i} are its decimals. Now, let us construct the number $0.b_1 b_2 b_3 \ldots$ the decimals so $b_1 \neq a_{11}$, $b_2 \neq a_{22}$ and so on, down the diagonal of this scheme. Such a number is real and different than any number in the list because it differs from each in at least one decimal place. Hence, you can't list all real numbers, you can't count them – there are uncountably many of them."

Lecture was over.

Lecture Two: Demons

"Angels and demons are a popular topic – but not a scientific one. To be precise, science does not talk about popular demons, but why wouldn't we imagine some creatures with supernatural abilities and place them in thought experiments?" After that introduction I resumed with the story:

"Two new guests weren't in a hurry. They were brought to the town by an unusual call: Laplace's demon's corpse was found in a train. "Laplace's demon? That's…", the older guest started the conversation. "Yes, that's Maxwell's demon's brother, my dear A.", the younger one answered, continuing: "You know how brothers have a special connection, like spooky action at a distance – as soon as one demon died, the other one felt it and reported the death of his brother, not seeing the corpse. Now I remembered, you have often sent twins on voyages yourself. Luckily, they always came back alive." That remark made A. laugh and say: "I know Maxwell's demon is a creature with a task – to open and close doors, but what about Laplace's one, what did he do?"

"He wasn't supposed to do, he was supposed to know. Based on the laws of Nature and the current state of the Universe, or any earlier state of it, to know everything about the world in present, past and future. Prediction is very difficult, especially about the future." the younger said and took the room key."

I didn't wait for the questions from the students, explanations began right away. "I know you've heard of both James Clerk Maxwell and Pierre-Simon Laplace, but this is probably the first time for you to hear about these creatures they made.

Maxwell's demon is an imaginary creature with a task, as A. said. The task is to defy the second law of thermodynamics. You know it already: heat transfers from the warmer to the cooler body, entropy of the system does not decrease. What does Maxwell suggest?

Imagine two tanks filled with gas, each at a different temperature. What does it really mean? One tank's kinetic energy of the particles (and hence the velocity) is smaller on average, and in the other one it's larger. However, we must not forget that still in both tanks exist slower and faster particles. If we took the slower particles from the warmer tank and replaced them with faster ones from the slower tank, result would be heating up the warm tank and cooling down the cold tank even more.

The task of this supernatural being imagined by Maxwell (and named 'demon' by Lord Kelvin) is exactly that: letting faster particles from the cold tank in the warm one and slower ones from the warm tank to the cold tank. For that task, we imagine the tanks are connected with tiny door controlled by the demon. He waits for the particles he needs to let in the other tank come close to the door and opens them. After enough time, he will finish his task and bring the system in the aforementioned state: heat will transfer from cooler to the warmer body, reducing entropy and breaking the second law of thermodynamics."

"But no one ever told us second law of thermodynamics can be broken", one of the students said out loud.

"I know! One of the following lectures will cover this in detail. Until then, think about possible answers to this apparently paradoxical question we opened right now" I said, and added: "You've heard of perpetuum mobile?"

"The device producing useful work without addition of energy?" Misha asked.

"That's perpetuum mobile of the first kind – and it breaks the first law of thermodynamics: the law of energy conservation. As energy cannot be created out of nothing, such a device is impossible. However, there is perpetuum mobile of the second kind, device converting heat to work spontaneously."

There I got interrupted by one of the students, Fedja: "We have those machines, what's the problem? They do not break the first law of thermodynamics – energy is not being created, just transformed to another form."

My answer was brief. "There are devices converting heat to work, but it is always the transition of heat from warmer to cooler body and you can never convert all heat to work. Perpetuum mobile of the second kind would be able to convert all heat to work, which is impossible. It would ask for only one tank which would be cooling all the time, which would eventually ask for heat transmission from the cooler to the warmer body, breaking the second law of thermodynamics. So, our demon is a perpetuum mobile!"

"Speaking of devices called perpetuum mobile, there's one thing I don't understand", Fedja spoke again. "Perpetuum mobile means perpetual motion – but isn't that what a ball on a flat surface does without addition of energy? Inertia?"

"It is, and that's sometimes called perpetuum mobile of the third kind. It is impossible as we cannot remove friction totally, and compared to the perpetuum mobile of the first and second kind, it cannot serve as an infinite source of work, since any removal of energy from the ball would slow it down and stop it, right?"

After this thermodynamics lesson (whose length I wasn't satisfied at all – I wanted to talk more, explain it better, I just couldn't find the right words), I made a break.

Then I continued. "Laplace's demon wasn't named by his creator either, he got the name later. Laplace himself says: "We may regard the present state of the universe as the effect of its past and the cause of its future. An intellect which at a certain moment would know all forces that set nature in motion, and all positions of all items of which nature is composed, if this intellect were also vast enough to submit these data to analysis, it would embrace in a single formula the movements of the greatest bodies of the universe and those of the tiniest atom; for such an intellect nothing would be uncertain and the future just like the past would be present before its eyes."

Imagine such a powerful creature! It knows what number will the dice show, because they obey Newton's laws. It knows where each electron is, where will it be… Scary, right? I'd have a lot to add about him, but I'd better keep quiet. The tread will unwrap in the lectures to come."

"Remember Borges' Irene Funes?" Mira asked. "Man of extraordinary memory, who always knew what time is it, without a clock… your demon!"

I nodded, amazed by the connection I didn't see before.

"Professor, you still haven't explained the part about twins, voyages, something you called spooky action at a distance… what's all that about?"

"I'll leave those questions unanswered today, since they are the foundation of the next lecture. I'm happy that you're concentrated and want to know it all right now, well done!

However, before ending this lecture, I would like to mention the third demon, the one not mentioned in the story, the scariest one. It's scary because it's real and present in the world: Morton's demon.

The creature was discovered – I deliberately say discovered to emphasise the cruel reality – by Glenn Morton in the creationist environment, but its effects are visible outside that specific group of people as well. This creature just like Maxwell's demon opens and closes doors: in people's minds. It allows entry only to arguments supporting their beliefs. For all other arguments, the door is firmly closed, protecting the demon's protege from any doubt. Demon delivers damage to everyone – to its protege, by letting him sink deeper and deeper in wrong theories and trying to explain them to the world, confidently and stubbornly – but also to those trying to give evidence against the evil victim's theory."

"Unfortunately, it easily occupies people's minds", Mira continued, "especially those who take some idee fixe to fit their own prejudices, desires and fears in them. These followers become more dangerous than the creators of the ideas they have taken, convinced that them and only them are right. I could accept that if their actions were just proselytism and internet troll fights. But when they bring someone's life to danger, when their radicalism result in someone's death… I cannot forgive. I can forgive them, because they are the victims as well – but I can't forgive the demon."

Later I discovered this was personal experience speaking.

Lecture Three: Einstein

"Time to explain the missing parts from the previous lecture… but first, a new paragraph. Please note, though – today's lecture is going to be tiresome. You could feel confused, frustrated or extremely curious. I can't cure all those symptoms on such a short notice, so you'll have to cure it with additional literature – it will surely be useful." After this introduction, I began with today's reading:

"And why did they call us, N.?" A. asked. "You were called because the train the corpse was found in was your idea all along, and I was called to confirm a doubt they had."

A: "What does Einstein Express have to do with anything?"

N: "Probably nothing – but it's good to have you here."

A: "We'll solve it easier together. You're lucky this didn't happen in Supplee's submarine or Bell's space ship."

"Does this mean A. is Albert Einstein?" Fedja was surprised.

I nodded, pleased: "Yes, it does. Maybe it's obvious who are the twins we mentioned last time?"

"It is," Misha confirmed. "It's the thought experiment in which one brother stays on Earth, and the other travels the space in an enormously fast spacecraft. When he is back, we see that the traveler is younger than the brother left on Earth, as a consequence of Einstein's relativity theory, time dilatation."

"Great, but… why does it happen? From the Earth perspective, the one in the spacecraft moves and time goes slower for him, as Einstein says. However, looking from the craft, the one on Earth is moving in the opposite direction, so according to that symmetry they cannot have different age!"

"No! There is no symmetry, the brother in the spacecraft has to make a halfturn to Earth, getting the acceleration and deceleration the one on Earth doesn't get – and that's where you get the difference and lose all the symmetry," Fedja reasoned brilliantly.

"Borges' Jaromir Hladnik is given the blessing of one second in front of the shooting squad lasting for a year", Mira added silently.

"Now, you can go in deeper analysis of these things, and you surely will do it later, but we'll stop right here and go for the next experiment: Einstein Express.

Imagine an extremely fast and extremely long train passing the place where Fedja stands. Misha is inside and stands in the middle of the train. When Misha is right next to the place where Fedja stands, train is hit by two lightnings, one at the front and one at the back, so the strikes are simultaneous for Fedja. Are they simultaneous for Misha as well?

No! Light got to Fedja at the same time because he is not moving, but Misha is – so he'll be reached by the light of the lightning in front before the one hitting in the back. That is why Misha doesn't think they've hit simultaneously."

Some students nodded, some shook their heads… but I couldn't stop, I was in a fast train.

"If this sounds weird, believe me, things can get weirder. N. mentioned Bell's spaceship – and that's another thought experiment about relativity of simultaneity like the Einstein Express – what happened at the same time for Fedja didn't happen at the same time for Misha!

Let's assume we have two spaceships connected with a tight rope, and that they start accelerating in the initial moment in such a way that for us, the observers from Earth, they keep the same distance. Is the rope going to break?"

"Of course not, the distance does not change!" a natural Newtonian conclusion came from one of the students.

"It doesn't change in our inertial reference system – but what about the system the ships are in? You know, Einstein's theory is asking and answering the questions how different motions look in different reference systems.

Is the distance between them going to increase, and the rope hence break?

Is the distance going to stay the same, but the rope shrink due to high velocity and break?

Is there a possibility for the rope not to break?"

After these questions students were loud – everyone had her or his own answer. I didn't have one, though: "I do not plan to tell you the answer, as the discussion is still heated – have a peek in the literature and find answers for yourself, analyse, question – questioning is not a sin."

I had to leave them confused and unsatisfied once again and proceed with new questions and paradoxes – next one was the submarine.

"N. also mentioned Supplee's submarine, so we have to say a bit about it as well. If you place a body with same density as the water in water, what will happen?"

"It won't rise, it won't sink, it will stay where it is" Misha answered, happy to finally get a question from classical physics.

"Good. Now, imagine a submarine with the density equal to the density of water and moving very fast under the water surface. What does the Einstein theory say? Fast moving body's density grows, so the submarine should sink, right? However, as the submarine moves observed from the outside, so does the water around it move observed from the inside, so the density of the water rises – hence the submarine should float out!"

"You know already – we have no clue" Misha was mildly depressed.

"First thing to notice here is the effect of gravity – and you don't have gravity in the special relativity theory, so we have to call the general one to help. General relativity gives an answer – the field collects energy at high speeds and attracts the submarine, so it sinks.

However, Supplee himself offered a solution in special relativity theory frame replacing gravity with an accelerated movement of the sea floor upwards – and there he managed to prove the imminent collision of the submarine and the floor based on the curvature of the floor.

I am quite sorry we have no time to talk about non-euclidean geometry, the parallel postulate, Lobachevsky, father and son Bolyai… But you know where the library is. Before Mira jumps in… yes, I know that Borges said there is a Greek labyrinth consisted of a single straight line a lot of philosophers got lost in – and that's every line in the curved spacetime."

"Just to say, we're exhausted, dizzy, and disoriented." Misha said.

"Great! That's motion sickness – we have changed a few means of transport, starships, submarines, trains, so nothing to wonder about there. Now we just have to solve the question of spooky action at a distance."

"Can we finish it next time?" Fedja was devastated.

"Sure, but then I have to give you another thought experiment to think about!

You're putting a ladder in your garage at an enormous speed. Because the garage is small and the ladder is too long, you hope the relativity effect will shorten the ladder and be able to fit it in the garage. However, from your perspective the garage is getting smaller, so you can't place the ladder in again. What happens, which interpretation is correct?"

Lecture Four: Chaos

The last lecture on Einstein was quite hard, so I decided to simplify this one, but make it interesting as well.

As the students expected, I started with a paragraph from the story.

"There was no need for the two of them to see the crime scene. In this world of thought experiments it was enough to think and talk.

A: "What can kill a person who knows it all?"

N: "Something he doesn't know: maybe the mere fact he can't know it all."

A: "Do you think it was Lorenz' butterfly? Flapped its wings somewhere in the rainforest, caused a storm, storm wrecked a home, home owner went crazy and killed the demon?"

N: "Interesting... however, Lorenz chaos, no matter how sensitive to change in initial conditions, is still deterministic – and Laplace's demon lived thanks to determinism. So, the butterfly is innocent, and so is Tesla's ray of light."

I was angry as I read this part of the story – it seemed as if I tried to make it as convoluted as possible. That's why I hurried with explanations, to avoid confusion: "The word chaos is usually used when we think of disorder, something unpredictable and not complying to known laws. However, the discipline popularly known as chaos theory, deals with systems that comply to clearly defined and known physical laws, but thanks to certain special properties they have unpredictable and unusual dynamics.

At the turn of the century, in 1900s Henri Poincare was investigating the so-called three body problem: the problem of determining trajectories of three bodies in a gravitational interaction – you can imagine three planets far far from all other celestial objects.... or Sun, Moon and the Earth. Two body problem is rather simple, so one would expect that the three body problem could be solved exactly. However, Poincare has shown that the nonlinear differential equations describing this system cannot be solved analytically to get explicit functions as solutions. The surprising part was seeing that there are trajectories that are neither periodic, convergent or divergent.

Ed Lorenz used a computer back in the early sixties (and you know how computers looked like back then) to analyse a simplified climate model. One day he wanted to repeat a simulation he made the day before, but not from the beginning, he was just interested in a part of it. This is why he took the values of system variables from the previous simulation at the time point interesting for him, and used them as initial conditions of the new simulation. However, the simulation results were nowhere near those obtained the first time. Why? The only difference was the fact that Lorenz used rounded values for initial conditions, instead of exact values from the system calculations. Could the tiny rounding error cause such a grave difference? Shouldn't the system have similar behaviour for similar initial conditions?

That is the characterisation of chaos: high sensitivity to initial conditions, and trajectories neither periodic nor stable nor unstable."

"Oh, that sounds like that die, demon knows the final outcome as he knows the initial conditions and the law it rolls!" Misha concluded.

"Analogy is approximately correct, the only difference is that you don't need the infinite precision for the dice, while in case of chaos you need it for the initial conditions – as any rounding leads to different trajectories very quickly. Maybe chaos didn't kill the demon, but it surely made his existence a lot harder" I answered and returned to the story: "Lorenz found a good example of this phenomenon with a question 'Can a flap of butterfly's wings in Amazon rainforest cause a hurricane in Texas?' That is the butterfly A. mentioned in the story."

"What is Tesla's ray of light mentioned in the story? Is it Tesla's secret weapon?" Fedja was curious.

I couldn't help but laugh: "Oh, you're reading our popular "science" authors! Those stories of Tesla's secret weapon, conspiracies and whatever else seem to have a good market out there. However, this is actually a reference to a quote by Tesla speaking of the butterfly effect before Lorenz: "A single ray of light from a distant star falling upon the eye of a tyrant in bygone times may have altered the course of his life, may have changed the destiny of nations, may have transformed the surface of the globe, so intricate, so inconceivably complex are the processes in Nature.""

Mira added: "Just like Borges says in The lottery of Babylon: "There are also impersonal drawings, whose purpose is unclear. One drawing decrees that a sapphire from Taprobana be thrown into the waters of the Euphrates; another, that a bird be released from the top of a certain tower; another, that every hundred years a grain of sand be added to (or taken from) the countless grains of sand on a certain beach. Sometimes, the consequences are terrible.""

"Now you reminded me of Kolmogorov's sentence that "the human brain is incapable of understanding really complex things" – everything we understand looks simple", Fedja commented.

"You know, complexity is today a science of its own. Right after the work Lorenz did, a lot of systems in nature and society were shown to behave in the same manner as Lorenz' simple chaotic system: numerous examples from fluid mechanics, lasers, electrical circuits, economy, biology, medicine show the chaotic behaviour appearing in the same way, sets of similar trajectories emerging with a non-integer dimension…"

"Like fractals?" Misha asked.

"Yes, just like fractals. Just like we found patterns of chaotic behaviour in different scientific areas, we also found patterns of incredibly complex behaviour which was seen in the systems traditionally considered as random as an oasis of order, while in the systems usually considered perfectly ordered and stable – and oasis of disorder. Those dealing with complex systems called them the edge of chaos, a perfectly fitting name.

I am so sorry we can't stay on the topic of complex and chaotic systems, so active fields of scientific work and research – in this course, as you might have noticed, we can't stick to anything too long: giving a few pointers to tickle your imagination and off we go. That's the quality of it, we're not opening doors for you, but we're pointing at some that you might not be aware of – and your scientific future could be behind them."

"Speaking of doors," Misha jumped in, "can you explain a problem we had a lot of discussion about in the probability course?

Imagine you're a contestant in a TV show with the following rules: at the beginning, three closed doors are in front of us, and the host informs us that behind one of them is the prize – a car, and behind the other two is nothing (or a goat, if we want to make it humorous). We choose a door and the host opens one of the two other doors and shows there is nothing behind them – he can always select the door with nothing behind them because he knows beforehand what is behind which door. After that, he asks us do we want to switch the door we've chosen with the other door left? What is better, switch or not to switch? Most people say it doesn't matter, but I've heard the correct answer is to switch?"

"Question isn't really related to our story – but we can always call it a thought experiment. It's the so-called Monty Hall problem, named after the TV show host – and yes, the correct answer is to switch." I answered, knowing Misha will not be happy with a short answer, though.

"But why? What is the difference?" Misha asked.

"Imagine for a second that the game is a bit different: the number of doors grows from 3 to 100 – and still you've got a car behind exactly one of them. The host opens 98 doors with nothing behind them and leaves you with a choice: stick to your first choice or switch. Think about it: the chance you've picked the right door right away was mere 1%. So in 1% your door will be the right one, and in 99% it's the other one."

"That's right."

"So, it's in our best interest to switch. Now if we look back at the original problem – we had a chance of 33.33… % to pick the right door, in the other 66.66… % it will be the other one, hence it's better to switch!"

"Oh, now it's clear," Misha confirmed and added a question: "but why does this result sound so paradoxical?"

"Because we are forgetting the game show rules: number one, that Monty knows where the car is and that he'll always open a door with nothing behind, and second, that Monty is always offering the switch. Forgetting the first one affects our calculation of probability, because if Monty was opening a random door, then trade wouldn't make sense as in 33.33…% Monty would open the door with the car himself and eliminate the advantage we would have with the switch.

If we forget the second rule – i.e. the switch is offered each time, no exceptions – we allow the psychological effect of doubt: they are offering us a switch, maybe they don't want us to win the prize…

See you next week!"

I finished the lecture with those words and chose the only door in this show, the lecture hall door.

Lecture Five: Entropy

"A: "What about the brother's old illness, the entropy?"

N: "You are right, that cannot be dismissed as an option. But let's go on, what else?"

Before A. and N. find a new suspect, let's stick to the term entropy. What is the first definition you've heard?"

"Measure of disorder in a system," Misha was brief.

"Oh, just like Borges says in Lottery of Babylon, an intensification of chance, a periodic infusion of chaos into the cosmos…" Mira was… well, just Mira.

"That exact description is good to illustrate the nature of entropy – time just increases it in the system, it can't be lowered without external influence. Every reduction of entropy, which may correspond to e.g. reducing temperature somewhere asks for entropy increase somewhere else. How does it look like in the case of Maxwell's demon? He reduces the temperature of the colder and increases the temperature of the warmer tank, which is by the second law of thermodynamics impossible.

As Arthur Eddington put it, it's a lot more probable that those monkeys write all the works in British museum library than to have the entropy decrease spontaneously."

"So, where's the catch?!" I almost screamed to wake them up. "We have an obvious entropy decrease in the system!"

A student from the last row, Ana, was the first to speak up: "When we put a pile of stamps in a stamp album, we reduce the disorder, but we invest a lot of work in it."

"Which in case of Maxwell's demon means he made more entropy than he eliminated from the system, and since he's a part of the whole system, the net sum would show increase in entropy? Correct!" I was pleased by the reasoning shown. "Now, where did the entropy increase come from? Unfortunately, we don't have enough time to explain all the necessary concepts, but we'll try to simplify it. Everyone in need of more technicalities, the library is downstairs.

The basic reason for the entropy increase in case of Maxwell's demon is measurement of speed and position of particles and saving that data. Szilard calculated: if every measurement brings an increase of entropy of at least $k \ln 2$ where k is the Boltzmann constant, the demon will not be able to break the second law of thermodynamics.

(A student interrupted me here with an exclamation: oh, that's the great Boltzmann! Yes, yes... I almost forgot how much Boltzmann did for physics. Everything I was talking about in this lecture was a derivative of his theory of entropy.)

Brillouin went on working on Szilard's solution and determined that the observation of a molecule needed for measurement represented by a photon sent to it brings energy dissipation and entropy increase compensating everything the demon did bringing in the order.

However, Bennet suggested a way to detect molecules without using light. This would discard Brillouin's explanation, but some circles still consider measurement of speed without light impossible and hence it is not yet an accepted fact that all measurements can be conducted without entropy increase.

Bennet still claimed the demon cannot break the second law of thermodynamics thanks to the Landauer's principle: every logically irreversible manipulation of information (like erasing a bit) must be followed by an increase of entropy in the environment, and the minimum energy needed to change one bit of information is $k \ln 2$, which matches Szilard's calculation.

So, the problem is erasing the recorded data in demon's memory. No matter how big memory he has, he will have to come to end of it and start deleting old data. As the increase of entropy in deletion is not compensated in reduction of entropy in data, entropy increases. So, even if we accept the possibility of measurement without entropy increase through a thermodynamically reversible process, the limited memory results in entropy increase due to erasing which is an irreversible process – and the demon fails at breaking the second law of thermodynamics, theory has survived the attack of supernatural."

"I've got two questions. What would data entropy mean and what does it have to do with the other demon?" Misha asked.

"First question is answered by a concept known as information entropy: measure of uncertainty related to a random variable. Larger the entropy, less is the chance we will guess the state or the variable value – it corresponds to increased disorder in the system. Concept was developed by Shannon, and von Neumann recognised it is an analogue of thermodynamical entropy in the world of information.

The analysis of Maxwell's demon has shown that, although essentially different, these two types of entropy are connected. Energy we get based on the particles passed through the door we use to form entries of information, trying to keep some sort of balance – but the extra cost of deletion and forgetting the particle state in the end destroys all balance.

In the other demon's case, the question of the huge number of calculations he has to do is also the question of the relationship between information and thermodynamical entropy – can the demon calculate everything in this universe using all the energy the universe can offer? That question is close to the already mentioned irreversible processes and imminent entropy increase – it has been shown that the demon doesn't manage those processes well: he is used to reversible ones."

I had to add this too: "Speaking of Eddington earlier, I have to tell you something very important. In the 1930s people it was thought that the fine structure constant α, the characteristic of electromagnetic interaction and one of the most important constant in the universe is approximately 1/136. Eddington made a series of arguments, mostly non-scientific ones that it is exactly 1/136, and he went one step ahead: that the number of protons/electrons in the universe is exactly 136×2^{256}. After a while, it has been demonstrated that the fine structure constant is closer to 1/137 than 1/136, and Eddington just tweaked his arguments to match the new value.

The Nobel prize winner Hans Bethe made a successful practical joke on Eddington's non-scientific principles and published the following paper in Die Naturwissenschaften journal:

(I took the original paper and read it)

Remarks on the quantum theory of the absolute zero of temperature
by G. Beck, H. Bethe, and W. Riezler

Let us consider a hexagonal crystal lattice. The absolute zero temperature is characterized by the condition that all degrees of freedom are frozen. That means that all inner movements of the lattice cease. This of course does not hold for an electron on a Bohr orbital. According to Eddington, each electron has $1/\alpha$ degrees of freedom, where alpha is the Sommerfeld fine structure constant. Beside the electrons, the crystal contains only protons for which the number of degrees of freedom is the same since, according to Dirac, the proton can be viewed as a hole in the electron gas. To obtain absolute zero temperature we therefore have to remove from the substance $2/\alpha - 1$ degrees of freedom per neutron. (The crystal as a whole is supposed to be electrically neutral; 1 neutron = 1 electron + 1 proton. One degree of freedom remains because of the orbital movement.)

For the absolute zero temperature we therefore obtain $T_0 = -(2/\alpha - 1)$degrees.

If we take $T_0 = -273$ we obtain for $1/\alpha$ the value of 137 which agrees within limits with the number obtained by an entirely different method. It can be shown easily that this result is independent of the choice of crystal structure.

It is obvious just how many things in this paper are wrong: the choice of using degrees Celsius, the rounded value of absolute zero… Authors did a great job at imitating scientists blindly holding to an idea and adapting everything to them: as if the Morton's demon got them.

Today we know the denominator in the fine structure constant is a bit larger than 137, isn't an integer nor there's a reason why it should be an integer.

Lesson: that's numerology, Max!"

Misha had something to add, though: "I'm worried about the fact that a satirical paper could have been published in a serious journal… What's with editors and reviewers?"

"Since you mentioned it, I don't know if you remember Sokal affair: the physicist who published a nonsensical and deliberately ironical paper on physics and social sciences in an eminent social science journal to show the lack of criteria and careful peer review in the area. He has shown the editors are ready to publish a paper they do not understand if a person with respectable references signs it.

Rematch was the Bogdanov affair, or as John Baez called it, anti-Sokal affair: two brothers published series of papers without sense – imaginary ideas framed in scientific jargon. Their Ph.D. dissertations had the same flavour. All in all, both cases show that we have a significant problem in review – the reviewers often see no interest in it: it's not financed and it's tedious. Because of that, we have Bogdanovs, Sokals, but also some bad papers that are just that – bad papers."

Misha added something again: "Isn't plagiarising worse?"

"Here's an anecdote from the beginning of my career about that… Maybe you'll think it's senseless, but I retell it every time hoping someone will explain how did my colleague guess where my family is from.

I was at a conference somewhere in the Ural and I was just telling a colleague from Dnipropetrovsk about an interesting event – I had a chance to browse a book by a colleague mathematician in whose index I found name of my father. However, the page pointed at by the index contained no mention of him.

"Your family must be from Vladivostok?" the fellow from Dnipropetrovsk asked.

"Yes, how did you know?"

"Intuition" he smiled and continued: "That author can never surprise you, Pravda and Izvyestiya are writing about his plagiarism of my own work for 6 months now."

Although that doesn't seem to be the most logical point made in this anecdote – I am retelling it fascinated the man knew my family was from

Vladivostok, as my best friend doesn't know it – plagiarism is the worst way of promotion and career development. You won't harm science, you'll harm yourself and people around you."

Lecture Six: The Sprinkler

After a tough lecture like the last one, it was my pleasure to announce a lighter part of the story. "In order not to think N. and A. are some dark, always serious people, today's lecture will show how does a joke a la N&A look like."

After this introduction I read only one sentence from the story: *"A: Maybe Feynman drove him mad, saying one day that the sprinkler is rotating clockwise, and the other day the opposite? Or maybe he was worried about the blue eyed islanders?"*

Then I've put a common garden sprinkler on the table and asked: "What happens with the sprinkler when the pump starts working and pushing water through it?"

"It rotates – as if powered by jet engines" Fedja said.

"Excellent description, thank you! Now, what if we put the pump into a pool full of water and make the pump work in reverse, i.e. suck water through it? How would the sprinkler move now?" I asked.

"Reversed with respect to the earlier state. It will go towards the water coming in. Water pressure is lower for the water coming in than the water on the other side, so because of the difference in pressures, water goes along the resulting positive pressure, which is opposite to the movement we had in the first case", Fedja gave his reasoning.

Another student, Alyosha, didn't agree: "Not necessarily! You're forgetting the momentum from the moving water hitting the inside wall of the sprinkler. It is just like the one caused by flow coming out of a standard sprinkler. Hence, sprinkler will rotate just like it would in regular mode."

Misha, sitting between the two, seemed completely confused: "While Fedja was speaking, I was sure it was rotating the way he was saying. When Alyosha spoke, the sprinkler in my head just changed the direction. I don't know what to think anymore!"

"Exactly what one of Feynman's professors said when young Richard convinced him in one direction one day, and the other the other day. As Feynman says, all physicists he ever discussed this problem thought the answer was obvious. Half of them thought it was obviously clockwise, while the other half thought it is obviously counter clockwise. By the way, the problem originated from Mach's mechanics textbook, and new literature brings more analysis on different level – with different conclusions as well."

After these words I made a dramatic pause (knowing everyone is just waiting to hear the correct answer, who is right?) so I continued: "Feynman decided to experimentally test the conclusions obtained through analysis, so he designed an experiment – in the cyclotron laboratory at Princeton. Experiment ended with a big bang – CERN guys aren't the first people making the big bang in particle accelerator labs. Feynman kept increasing the pressure during the experiment which led to the explosion of the bottle used, with glass and water flying around."

"Well, did anyone repeat the experiment? What is the correct answer?" Misha was impatient.

I was ready for this. "There were experiments, calculations, discussions. If we are thinking of the idealised sprinkler model, answer is between Alyosha's and Fedja's: the sprinkler isn't rotating at all. The two effects you mentioned simply cancel out, since their intensity is the same according to the calculations."

However, one word seemed suspicious in my answer: the idealised model. Misha noticed it: "What if there is fluid viscosity, what if there is turbulence?"

After a moment, I answered that as well: "It would reduce the momentum of the flow as it enters the sprinkler, the one Alyosha was talking about. That is why Fedja's effect would dominate and the sprinkler would go towards the flow, not fast but it would probably move. However, you know I'm a theorist, and real concepts don't impress me much.

If today's story was interesting, play around with some maths, try to calculate the important quantities yourself."

"What does this have to do with the big story, though? The demon had to know the answer to this question, as the system is completely deterministic, everything is clear, right?" Alyosha asked.

"Correct, this wouldn't be a problem for the demon. Notice that A. wasn't serious when he mentioned the sprinkler as a possibility, he just wanted to remind N. and himself of this story. You learned something new thanks to it, didn't you?"

"And the blue eyes, the island", Misha asked.

Oh, I did forget about that part. "Thank you for the reminder, we almost skipped the explanation of that. So, just like Feynman was able to convince people around him in both of the two potential outcomes of the sprinkler experiment, I was able to confuse people with the two possible outcomes of the following story: On an island in the middle of an ocean lives a tribe with a strict rule – no one may know what is his or her eye colour. That is why they do not look in the mirror or ask others to tell them. If they find out the colour of their eyes, they would have to commit a ritual suicide on the town square the next day at noon. The island is populated by n blue eyed people and m people with other eye colours.

One day, a storm left a ship on the island's coast. The islanders helped the captain repair it and he was very grateful for the help indeed. When the ship was ready, the whole island gathered to say goodbye to the captain. While sailing off the coast he thanked for everything and expressed his wonder that on an island in the middle of the ocean he saw people with blue eyes, like his. After those words, his ship was already far, far away.

The question is: what will happen with the islanders? Assume they are perfect logicians."

"It depends. If the number of blue eyed islanders is one, that poor fellow will come to the conclusion he never saw a man with eyes like captain's, and hence he'll know his eyes are blue (as the captain said there are people with eyes just like his among them). Next day he'll kill himself", Misha started unwrapping the mystery after a short pause.

"Correct. But what if there's more than one blue eyed islander?", I asked.

"Nothing, because they knew there are blue eyed people among them. Captain didn't bring any news, hence – there is no reason for anything to happen" Misha answered.

"I don't agree", Alyosha jumped in. "If there are two blue eyed islanders, each one will assume the other one is the only blue eyed person on the island. As the other person doesn't kill himself or herself the next day, each one will know that he or she has blue eyes as well and commit suicide the next day. If there are three of them, each one will see two blue eyed people and expect them to commit suicide after two days, according to this scenario. Once they don't, they'll all commit suicide on the third day. Inductively, if n blue eyed people were in the island, n-th day they all die."

Misha wasn't giving up: "But why did they need the captain to tell them there are blue eyed people among them? They knew it already. Since they knew it, why didn't this procedure start before he came to the island. If it did start before, when did it start?"

I was pleased to announce: "You got to the two solutions of the puzzle I had in mind. They both sound pretty logical, but can't be true simultaneously. Those agreeing with Alyosha, please raise your hand. And those agreeing with Misha?" Result was roughly 50-50, as I expected.

"Not to keep you in suspense, I'll tell you right away – Alyosha is right. However, Misha's question is the key: what is the news the captain brought to the islanders, if there was more than one blue eyed man there? Answer might be a bit unclear, but think about it.

Assume there were 5 people with blue eyes, call them A, B, C, D and E. Put yourself in the position of person A and think: I see 4 blue eyed people,

B, C, D and E. If I am not blue eyed, then every one of them sees 3 blue eyed people, say B sees C, D and E. Then B thinks just like I do and says – if I am not blue eyed, then every one of the three blue eyed people I see sees two blue eyed people, C sees D and E. Why wouldn't C think as A and B then and think that if he is not blue eyed, each one of the remaining two see one blue eyed person each, D sees E and vice versa. It is logical to assume D thinks in an analogous manner as well, so D thinks that if he is not blue eyed, E sees no one with blue eyes. Since no one told him or her such eyes existed before the captain came, he enjoyed the bliss of ignorance until then."

"Wa… wait a second. E sees A, B, C and D!" Misha was confused.

I nodded. "Yes, he or she sees it. However, the argument is in the chain of thought: A thinks that B thinks that C thinks that D thinks that E may not know that blue eyed people exist. That is what keeps the island at peace before the captain came: after that, A knows that B knows that C knows that D knows that E knows blue eyed people exist. Of course, this holds for an arbitrary number of blue eyed people, 5 was selected randomly. That's it!

Beauty of these problems is that they occupy the mind and make you think about them intensively the moment you hear them. Often you can't think of anything else, but luckily you almost never get two such problems at the same time."

"There are never two Zahirs!" Mira noted.

"Borges? Really? Is this a physics course or a Borges course?"

Lecture Seven: Uncertainty

"Earlier we start, earlier we'll finish. So, get to work!" After saying this, I began reading the next paragraph:

"N: I'd rather say uncertainty killed him, my dear A."

"A: Heisenberg? Speaking of him, where is he?"

"N: Both Heisenberg and Godel, all those people who shook our belief in true and absolute. Where he is? According to his theory, I can't tell that with certainty."

"A: "What did he tell you back then?"

"N: Doesn't matter. (N. said this in a sad voice)"

I breathed in deep and started with the explanations.

"It's time to introduce three new characters in the story: Werner Heisenberg, Alan Turing and Kurt Godel. In which order? Chronological will do.

First one: Heisenberg, Werner Heisenberg. His discovery made the foundation of quantum theory – the more precise our measurement of particle position, less precise is the measurement of its impulse. Why is it so? Imagine having an incredibly powerful microscope looking at a single electron.

Observation in this case means sending a photon and waiting for it to come back. If this photon has a small wavelength and big impulse, we can get the position of the electron very precisely, but the photon gives an unknown impulse to the electron, so we fail to know its own impulse. If it's a large wavelength and a small impulse, the situation is reversed.

It appears that a similar connection exists between time and energy – however such analogy would be indirect, as Landau put it: if we had such an uncertainty relationship, it would be false which could be shown by measuring energy very precisely and then looking at the clock.

However, it is clear what does this principle of uncertainty mean: short-lasting state cannot have a precisely determined quantity of energy. Tamm defined it more precisely."

"Speaking of Tamm, did you know that in those violent days of the revolution and civil war, Tamm was caught by some gang members in Ukraine and taken to their leader as a potential spy. Our hero said he was a maths teacher, and the leader then made a request: if the prisoner can show what is the error produced by neglecting all members of Maclaurin series starting from index n, he may go. Tamm did it. Useful mathematics!" Misha tossed an anecdote in.

Anecdote for an anecdote: Fedja added one more: "Sophus Lie was caught under the premise he was a German spy in Italy. Legend says he said he was a mathematician right away, just like Tamm did, and then the judge asked Lie to give a lecture and prove he is a mathematician. Lie did it, but judge wasn't impressed: allegedly he said that Lie can't be a good mathematician, since he understood the whole lecture."

"I haven't heard those stories, interesting! Speaking of classics – I already mentioned Dau, Lev Landau today: have you heard of his comprehensive maths and physics exams called the Theoretical Minimum? Very few students made through those exams Landau and Lifshitz made – what would you say if we had those here at our university?" I replied with a question.

"God no! Theoretical Minimum exams weren't regular university exams – students would study hard after acing regular exams just to get a chance to pass Teormin. If they made it, they'd be in Landau's kruzhok, and we all know how many academicians and Nobel laureates came from that circle. There is no university in the world giving that kind of diploma!" Misha answered readily.

"Well done, great answer! Now let us continue, as it's Kurt Godel time. If Heisenberg popped the classical physics bubble, then it was Godel to do the same in mathematics, losing our faith in absolute truth.

We'd love to have a full and consistent theory of anything. Full means that any claim in that theory can be proven or rejected using the axioms of the theory, and consistence means that there is no claim that can be proven while also proving its negation.

Godel has shown that this wish cannot be granted even in the case of elementary arithmetics, the grounds of mathematics!

And finally, it was Turing to bring the critical blow. The man who decided the course of World war two by breaking the German codes played a vital role in science before that. Still unborn computers and computer science were sentenced to the curse of undecidability, same nature as the one Godel shown in mathematics.

You had to hear Turing's name in the context of Turing's machine, an abstract device often used as a computer paradigm. Imagine we have a code or a program description for a Turing machine – or a computer. Can we always answer the question whether that program will ever terminate or go into an infinite loop? Answer is negative, that is why we say that the halting problem is undecidable.

How many problems, how many questions not even the Demon can answer…."

"How can we prove these claims, Godel's and Turing's in particular?" Fedja was curious.

"What did Banachiewicz say to Sierpinski? Cantor!" I answered with a riddle.

"What does A's question in the end mean?" Misha asked.

"It doesn't matter" I answered sadly, just like N. did.

Mira added, as she usually did: "As Borges' narrator in The lottery in Babylon says: I've known what Greeks didn't, the uncertainty. It seems we didn't know it either before 20th century!" Nice conclusion.

Lecture Eight: The Interpretations

"Good morning! Today's lecture is a logical sequel to the last one, with interesting plots and ideas. I'm sure you won't be Bohred." After that, I continued the story.

"After those words, N. smiled and continued: "You know how curiosity killed the cat? Now, I think the curiosity about the cat killed the demon!"

A: "Now you're talking about Schrodinger?"

N: "Of course. The poor demon died in confusion whether the cat is dead or alive."

A: "Now you're joking, right?"

N: "Dear A., you know what the crime scene investigation showed? Demon committed suicide."

A: "How? Why didn't you say it before?"

N: "I didn't want to mention it before we found a few potential reasons for his death. He killed himself with a shot from a gun connected to a device measuring photon spin. Quantum suicide."

A: "He wanted to know if many worlds interpretation was true or not."

N: "Maybe, and maybe he was just depressed – haunted by uncertainty."

"Now this sounds way too… creative" Misha said.

"I know. To understand all of it, we'll need some theory. Let's start with our previous lecture acquaintance, Werner Heisenberg. Those used to Laplace's determinism and classical newtonian mechanics it was impossible to accept that we cannot know both the position and the speed of a particle. Despite that, a new, quantum mechanics was formed with the uncertainty as a major element of it.

Niels Bohr, Werner Heisenberg and others made the so-called Copenhagen interpretation of quantum mechanics based on calculations – hence it's often described as 'shut up and calculate'. In its essence, it is a theory based on the following principles: system is described by its wave function, vector of states being calculated where those states are probabilistic. Those states are the subjective impression of the observer about the system.

If we are waiting for an electron to hit a screen, we do not know where exactly will it hit, so the wave function is purely probabilistic. However, once it hits, the probability of hitting any other point drops to zero and we know something concrete and precise – this is called wave function collapse.

Now, back to the story. The cat N. was joking about is the famous Schrodinger's cat, a thought experiment in which a cat is placed in a closed box and its life depends on the state of a small quantity of radioactive substance which may or may not (with the same probability) act in a radioactive decay. So, in 50% of cases the decay will happen, and after it the Geiger counter in the box will register radioactivity and break a bottle with poison which will kill the cat. In 50% of cases there is no radioactivity, the bottle stays intact and cat stays alive. Looking at the box we can't say if the cat is dead or alive, its current state is the superposition of those two equally probable conditions. After opening the box – which is equivalent to measurement in quantum mechanics – the wave function collapses and instead of the probabilistic state, we get the real state.

But what if we put a man in the box together with the cat? Then we think the cat is in a superposition of states, but the man inside knows exactly which state is true. How can a cat have two wave functions?"

"Simple – you said the state is a subjective impression of the observer. We have our impression, the man in the box has his", Misha reasoned as a true Copenhagen theorist.

"Well done! Now, the story about Copenhagen interpretation ends with the question of bras and kets. This story is something that you'll hear again in your quantum theory course, but it's worth hearing now as well.

It's the notation introduced by Paul Dirac, and we all love it, some only because of the names. It is customary to denote the inner product with $\langle x, y \rangle$. If we put a bar in between we get $\langle x | y \rangle$. Now, Dirac says: $\langle x |$ is bra, $| y \rangle$ is ket. So, ket is a column vector and bra is a row vector of states."

"Cute names, I have some puns in mind," Alyosha laughed.

"I know, you're not the only one," I replied and continued: "As you might have guessed, Copenhagen interpretation is not the only one. There is a whole list of ways to interpret phenomena of quantum mechanics, and probably the most interesting for you will be the many worlds interpretation.

Unlike Copenhagen interpretation supporters, supporters of the many world theory suggest that the cat is in one world alive, in another one is dead. When the box is opened, there is no unification of the worlds in any manner: in one world, the observer sees a live cat, in the other a dead cat, no wave function collapse."

"Borges' Garden of forking paths!" Mira exclaimed.

"Yes. It's obvious that the two interpretations are quite different. Laplace's demon would surely want to know which one is the correct one. One way is the one proposed in the story:

We get inside the box instead of the cat and wait for the death. Let the revolver be connected to a device measuring photon spin, measuring a new photon's spin every 10 s. Depending on the value of the spin there are two equiprobable options: gun shoots, killing the person or it doesn't, makes only that evil click. According to Copenhagen interpretation, after the first photon click or the shot make the wave function collapse, so if the gun goes off, the experimenter is dead and all further experiment iterations change nothing.

However, in the many worlds interpretation, the first photon splits reality into two parallel realities, one in which the experimenter lives, one in which he isn't. The one with the live experimenter again branches into two and so on. Experimenter never dies in all realities.

This makes it clear – the suicidal experimenter will know which one is correct based on whether he survives!"

"You owe us something," Fedja coughed. "You never told us what is the spooky action at a distance."

"Excellent time to do so, thank you very much! It's a paradox named as EPR – Einstein, Podolsky, Rosen, a consequence of something Einstein called spukhafte Fernwirkung – spooky action at a distance.

The authors wanted to point out something that didn't work in the Copenhagen interpretation – imagine two particles interacting and then being separated so the state of one is directly correlated with the state of the other. Measuring the sate of one we collapse the wave function for the other one as well, which could be light years away – which practically means that changing the spin of one particle we change the spin of the other. Does that mean information transfer which is faster than light? Einstein didn't like that.

Einstein hoped that this will show this quantum mechanics interpretation to be incomplete, while the other authors thought it's enough to add hidden variables to clear up the misunderstanding.

Copenhagen interpretation does say the wave function collapses instantaneously, but the observer of the other particle cannot profit from that information until he or she receives it directly from the first observer, which is necessarily slower than the speed of light. That's still something Einstein doesn't like, as it breaks the concept of locality, that only events in particle's neighbourhood affect it.

Bell's work has shown that even hidden variables cannot help here, and that no theory of local hidden variables can reproduce all what quantum mechanics expects. So, you get to a conclusion that the Copenhagen interpretation is simply non-local and as such it exists and gives correct predictions. On the other hand, the many world interpretation has no problem with locality, as there's not wave function collapse in the interpretation."

Knowing that any additional explanation will take us too far, I ended the lecture.

Lecture Nine: The End

"How many times have you taken a book called Short course in this or that and left it shocked with the number of pages it had?" I asked the students half-jokingly, remembering my own books with that innocent title.

"Luckily, this course bears that attribute for a reason – the story ends today, with a brief lecture without a defined title" I commented before turning the page and reading the end of story:

A: *"So, the answer to the mystery was in quantum theory all along. That's why they called you. But why this hotel?"*

N: "I can't say about the hotel, but talking of quantum theory: answer to the most of mysteries is in that world of bras and kets, except for the mysteries solved by your own theory."

A: "They say my and your theory cannot be simultaneously true – and I don't mind that as long as they both reveal the world around us, day by day. However, I do mind that the answer in this case was lying in uncertainty." (pause) "God does not play dice."

N. laughed. "A., stop telling God what to do with His dice."

Questions came right after I finished. Fedja started: "They called N. because it had something to do with quantum theory. So, N. is some kind of an expert in quantum theory?"

"The answer, my friend, is blowing in the wind", I replied.

"It's blowing somewhere in the story", Alyosha nodded.

"When N. says your theory, he thinks of relativity, right – we already figured out A. was Albert Einstein" Fedja commented. "Using that logic, when A. says your theory, he probably thinks of quantum theory? I think I got that, but still there are open questions. What does the Hilbert hotel Grand have to do with everything, is it just there because it's an interesting concept? Bras, kets, we already explained those… but why is it impossible for both theories to be simultaneously true? What is the whole dice story about?"

I was pleased with the questions and started leading the students towards the answers: "OK, let's go one by one. Hilbert hotel Grand is a concept closely related to infinite countable sets, as we already explained in the lecture number one. Which elegant approach to problems of proving uncountability did we mention?"

"Cantor's diagonal argument?"

"Correct. Recently, David Wolpert has used Cantor's diagonal argument to show the impossibility of Laplace's demon existence. Now you see another potential suspect in our story, don't you?" Not really waiting for an answer, I continued: "Next question was why can't both quantum theory and relativity be true at the same time? We mentioned one of the reasons the other day – EPR paradox. Action at a distance is not in spirit of general relativity. Why does A. say it doesn't matter, though?

Probably because quantum theory deals with the microcosmos and the relativity with macrocosmos, so they give satisfactory answers in their own domains. However, the humanity does want to unite those two theories and have a unique set of rules explaining physics. The missing ingredient is the theory of quantum gravity, this part of the puzzle still doesn't fit. Do you think it will fit eventually?

Stephen Hawking did think it will – but then he publicly announced his pessimism in the "Godel and the end of physics" lecture. However, Hawking makes a step further than N. and blames Godel not just for Laplace's demon death, but death of physics as well!"

I made a pause as the students laughed.

"Now the dice. I think I saw a hint that Alyosha knows what is it about. Am i right?" I asked the young student.

Alyosha was ready: "If I'm not mistaken, the first sentence is the quote by Albert Einstein, or A. as you call him, from a letter to Max Born?"

"Correct." I nodded. "And the second one?"

"Second one is Niels Bohr's alleged answer to Einstein, not to tell God what to do with His dice."

"Exactly so. So, it's clear now, A. and N. are Albert and Niels, and I'll let you think about cosmic dice and why Laplace's demon couldn't guess every outcome in that game of dice. I hope you liked the course, but that you haven't taken it too seriously.

My wish was to make you think about some questions. Classical courses give answers too – but this one just gave some hints. You've heard a few new terms, maybe poor and rigour-lacking explanations of those terms (I beg for forgiveness from mathematicians and physicists reading this) and that is enough. If you ever end up on this side of the lecture hall, try to give every generation that has the luck to be taught by you a course like this one. I mean – this type, not this quality. Make yours better."

I ended the lecture with these words and said bye to the students. Alyosha kindly asked me to copy my lecture notes… but I could only shrug and point at the two pages on the table – that's it. I promised to give a brief re-telling of the course in a story, because I don't see the sense of writing formal notes for an informal course.

What you are reading now are those promised notes. I didn't remember all details, and the notes are a few times shorter than the lectures themselves, but I'm sure they can help: those who listened to the lectures to remember some forgotten things, and those who haven't, to have a first look. Do you believe in love at first sight?

I complained to Mira after writing the manuscript: I became aware of how many similar – better – books exist about this topic and I wondered if there is any sense in giving it someone to read.

"If Pierre Menard could write Quixote again, you can write a book like this one…"

Sometimes I also hated the fact that the book is too short and too many explanations are missing. Mira again went with a Borges' quote:

"It is a laborious madness and an impoverishing one, the madness of composing vast books — setting out in five hundred pages an idea that can be perfectly related orally in five minutes. The better way to go about it is pretend that those books already exist, and offer a summary, a commentary on them."

Last question I asked her was if I put in too many details and too many puzzles I didn't solve in the story. Again it was Borges to answer.

"Mysteries ought to be simple. Remember Poe's purloined letter, remember Zangwill's locked room."

She didn't say the rest of the quote, so I did… finally satisfied with the book.

"Or complex. Remember the universe."

Part V

The Science Behind the Fiction

Afterthoughts

Most of the puzzles, allusions and homages can be directly understood from the text of the stories. However, there might be a need for explanation of a few things separately, in case the reader would like some extra clues. Consider this a walkthrough – and beware, spoilers ahead.

Normed Trek

Normed Trek is an Alice in Wonderland story under the cloak of Star Trek. I didn't realise it before I sat down to make an illustration for it (Fig. 1), exactly seven years after writing it.

The story offers an interesting problem: prove that $\sin nx / n \ln n$ is not a general term of a Fourier series (it was an exam question in my calculus class).

The idea that $|\chi(x) - 1/2|$ (where χ is the Dirichlet's everywhere discontinuous function) is constant also came from that class.

Finally, the expression of love in the last sentence asks you to do some drawing. Plot the implicitly given function $x^2 + 2 \left(y - \frac{3}{4} \sqrt{|x|} \right)^2 = 1$ and enjoy!

The Cantor Trilogy

All the mystery in the Cantor Trilogy is based on the fact that Molnar in Hungarian means Miller. Once you see that, hopefully it all makes sense.

It might be interesting to note that originally, JHM in the story stood for Journal of Human Mathematics. Only after my first contact with Gizem Karaali, editor of Journal of Humanistic Mathematics, have I realised that there is a journal with similar name: Journal of Humanistic Mathematics. I

© Springer International Publishing Switzerland 2016
H. Šiljak, *Murder on the Einstein Express and Other Stories*, Science and Fiction,
DOI 10.1007/978-3-319-29066-9_5

Fig. 1 The deployment of RAI in front of the Function and two Guards, Gamma functions with differentiators (spears with partial derivative tip) in a coordinate system (Image by the author)

edited the name and submitted the manuscript for publication in the Journal. Imagining anthropomorphic cantor scientists like the one in Fig. 2 was a fun thing to do.

In Search of Future Time

In Search of Future Time is essentially a short Arabian Nights piece. Most of the important explanations (including the explanation of Fig. 3) are given in Gadarine's letter (ciphered using Vigenere cipher, with Gadarine as the key), so they will not be repeated here. However, there are a few details I would like to address here.

Gadarine's surname Slavenski and the name of her great grandfather Egon is a homage to a brilliant young writer from Bosnia and Herzegovina, Karim Zaimović, as Egon Slavenski was one of his minor characters. Karim is my personal hero.

The first Gadarine's story is a dream. Rick couldn't read what the plaque said, it kept on changing: that's what happens in our dreams too. There are also three sheep metaphors: the Golden Fleece, the sheepskin coats and 'Arry S.

In loving memory of Abdus Salam Cantor, the first　　　　Nobel laureate for its work in physics.

Fig. 2 While making this illustration, the grave of the great physicist Abdus Salam was in my mind. Just like this Cantor named after him, he won the Nobel prize in physics. The text on his gravestone reads: "… in 1979 became the first XXXXX Nobel laureate for his work in physics." XXXXX denotes part of text that was removed later by the Pakistani government. Abdus Salam considered himself a Muslim, member of Ahmadiyya community, but Pakistan refuses to recognise the Ahmadiyya as Muslims. Hence, the text claiming he was the first Muslim Nobel laureate was edited by the authorities and the grave was defaced. The illustration here suggests that in case of Abdus Salam Cantor, the erased word is – computer (Image by the author)

Work of professor Komander in the next story is rather similar to what I've been doing for some time: making and assessing models which produce signals similar to human brain EEG signals, so it was inspired by my own dreams in a certain way.

Finally, a comment on the Internet of Things and People, mentioned in Pickering's (homage to Shaw's Pygmalion) letter: Internet of Things is becoming a common word today, so it wasn't too hard to take it a step ahead and put people in the same network.

Murder on the Einstein Express

All characters in this story were inspired by real people (or demons, as shown in Fig. 4), but only two are famous: Vladimir and Sergey. Names and surnames are shuffled, and Novikov (New) is changed to Starikov (Old), but the truth is that I had brilliant Russian mathematicians Vladimir Arnold and Sergey Novikov in my head while creating the characters of two professors.

Fig. 3 The wall without five paintings Gadarine took (Image by the author)

Fig. 4 I used an old graphic of Friedrich Engels writing under a portrait of Karl Marx. Here, Maxwell's demon is under James Clerk Maxwell's portrait, and he is opening a door on a box. The Greek vase behind him depicts a well-known scene: the birth of Pandora (Image by the author)

Jorge Luis Borges simply had to be in this story. Just like aforementioned Karim Zaimović, Borges is the literary inspiration I keep coming back to, ever since I read every single word he wrote during my freshman year in high school.

The story of Seryozha's family in Vladivostok and the man from Dnipropetrovsk is a homage to Tom Lehrer and a reference to his song Lobachevsky. There, Lehrer's alter ego claims that he copied the index in his book from an old Vladivostok telephone directory (and also that he stole a major discovery from a man in Dnipropetrovsk).

You might remember the question "What did he tell you back then?" A. asked. It was a reference to the famous Bohr-Heisenberg meeting in September 1941 in Copenhagen. The question of what was said in that meeting is a everlasting one, as the meeting itself was highly controversial. If you have a chance, go see "Copenhagen", Michael Frayn's brilliant play about this event.

In the end, I hope you saw the Bob Dylan reference! "The answer, my friend, is blowing in the wind."

Printed in the United States
By Bookmasters